Postphenomenology and Technoscience

SUNY series in the Philosophy of the Social Sciences

Lenore Langsdorf, editor

Postphenomenology and Technoscience

The Peking University Lectures

Don Ihde

Published by
State University of New York Press, Albany

© 2009 State University of New York

All rights reserved

Printed in the United States of America

No part of this book may be used or reproduced in any manner whatsoever without written permission. No part of this book may be stored in a retrieval system or transmitted in any form or by any means including electronic, electrostatic, magnetic tape, mechanical, photocopying, recording, or otherwise without the prior permission in writing of the publisher.

For information, contact State University of New York Press, Albany, NY
www.sunypress.edu

Production by Eileen Meehan
Marketing by Fran Keneston

Library of Congress Cataloging-in-Publication Data

Ihde, Don, 1934–
Postphenomenology and technoscience : the Peking University lectures.
p. cm. — (SUNY series in the philosophy of the social sciences)
Includes bibliographical references and index.
ISBN 978-1-4384-2621-1 (hardcover : alk. paper)
ISBN 978-1-4384-2622-8 (pbk. : alk. paper)
1. Phenomenology. 2. Postmodernism. 3. Philosophy, Modern—20th century. 4. Technology. I. Title.

B829.5.I345 2009
142'.7—dc22 2009004616

10 9 8 7 6 5 4 3 2 1

Contents

List of Figures

Chapter 1

Chapter 2

Chapter 3

Chapter Four

Introduction

When I first met Professor Jin Xiping during the summer of 2005, he was standing at the copy machine, hurrying to complete copies of materials impossible to obtain in China. I knew that he was a Husserl and Heidegger scholar, that he had studied for many years in Germany, and that he was the head of several of Peking University's philosophy institutes. During off moments, we began what was to be a much longer, discontinuous "conversation," as Richard Rorty would have called it.

Xiping knew of my technoscience seminar, and so we talked about the philosophy of technology. Inevitably the conversation would sometimes turn to Heidegger, the Heidegger of "the question" of technology. Xiping told me of the deep resonance many Chinese philosophers had for Heidegger, precisely for one of his traits that I had criticized: romanticism. This was not surprising, since the traditions of ancient China also held deep feelings of reverence for nature. It also was not surprising because I had encountered this nostalgia in many places, including my philosophy of technology travel venues, where I came to know many equally romantic Europeans.

But after further conversation, including some of my explanations as to why I thought this romanticism had led Heidegger to a dystopian and distorted philosophy of technology, Xiping did surprise me. He observed that as deeply as he felt empathy for the strand of nature romanticism that Heidegger exemplified, he knew that *such a philosophy of technology could not work in today's world*—and of course it could not work in today's China. We ended this string of the conversation when, with a second surprise, Xiping invited me to give a series of lectures on postphenomenology in relation to technologies at Peking University. That was the birth of *Postphenomenology and Technoscience: The Peking University Lectures.*

I had been to China only once before this conversation. In 2004, Professor Chen Fan from Northeastern University in Shengyang had invited Andrew Feenberg, Langdon Winner, and me to come to present a four-seminar series on the philosophy of technology. It was, collectively, our first trip to the potential superpower that is China. It also was the first time foreign philosophers of technology had been invited to Northeastern University. We went, the seminars were translated into Chinese, and the resultant publication (English title translation) was *Research in Philosophy of Technology in the Global Age* (2006). Our seminars were primarily about our own work in the philosophy of technology. The experiences of the trip were climax ones, ranging from the food—hundreds of dishes, virtually none familiar to us veteran Chinese restaurant gourmets—to the required steel mill visits, but also to the overall sense of the colossal dynamism of China's movement toward technological modernity. Is that why, one year later, something like postphenomenology trumps what I would have expected, a set of lectures on Euro-American philosophy in a "Continental" style of exposition and commentary? Or was there an echo of an earlier philosopher's visits? There was John Dewey's extended visit to China from 1919 to 1921. Dewey had gone to lecture on his philosophy, and in prerevolutionary China his pragmatic version of education and his notion of grassroots politics found receptive ears in early-twentieth-century China—but his liberalism did not. His student and host, Hu Shih, interpreted Deweyan pragmatism to find solutions to modernization in education and cultural reform (rather than in political action or revolution). Experienced China travelers find significant affinity within Chinese culture itself for pragmatism. My own connection—the adaptation of much of Deweyan pragmatism into phenomenology, now postphenomenology—I hoped would also strike that resonance nearly a century later.

The four lectures, now the four chapters of this book, were delivered in Beijing in April 2006 and are today published as a book in Chinese. My sponsors were professor Jin Xiping, one of whose titles is head of the Institute of Foreign Philosophy, and professor Wu Guosheng, head of the Institute for the Social Study of Science, both at Peking University. (The university retains its original name honoring its past, in spite of the more recent standardized spelling of its host city Beijing.)

In keeping with the common practice in China of having all lectures in languages other than Chinese available in advance, I had packets, complete with illustrations, printed up for participants. Participants of Stony Brook's technoscience research seminar also were given copies. Several veteran seminar participants pointed out that the lectures were a good summary of much of the last decade of development in postphenomenology and would constitute a good introduction to this style of philosophy. This is one motivation for an English version of the Peking University Lectures. There was a clear parallel between a new audience from another culture and a new audience among English-speaking readers new to postphenomenology and technoscience. The program is simple: Chapter 1, "What Is Postphenomenology?," describes and explains of how pragmatist threads and the empirical turn of science studies were incorporated into my use of contemporary phenomenology. Chapter 2, "Technoscience and Postphenomenology," traces a brief history of philosophy of technology as it developed toward the contemporary notion of *technoscience.* Chapter 3, "Visualizing the Invisible: Imaging Technologies," summarizes a decade of concrete research in imaging technologies, the "empirical turn" example of this book. Chapter 4, "Do Things Speak?: Material Hermeneutics," describes a parallel, but newer, research program that turns results from the previous program back to the humanities and social sciences.

The content of *Postphenomenology and Technoscience: The Peking University Lectures* is original in this volume. The chapters, however, did draw from previous work, and a bibliography of some of that earlier work appears at the end of this book. The earlier work was presented at a variety of venues—conferences, workshops, and seminars, primarily in Europe and North America, which had been the sites for my most frequent previous travels before the new century. China, along with Korea and Japan, has much more recently become a host to philosophers of technology, where there is much interest in this still-young discipline. Indeed, since Beijing, I made one more trip to China in 2007 to the Shanghai Academy of Social Sciences in Shanghai, hosted by Tong Shijung and Yu Xuanmeng, and then to South China University of Technology in Guanzhou, hosted by professor Wu Guolin. Once again the requested topics were entirely contemporary, with lectures on the state-of-the-art philosophy of science at the first venue and the same for the philosophy of technology at the second. Seminars were primarily focused on my own research programs, thus the pattern for all three visits has been contemporary and related to the technoscientific texture of the increasingly interconnected global world.

In developing this short monograph, I wish to acknowledge, in addition to my Chinese hosts and auditors, the very helpful editing suggestions by Lenore Langsdorf, the corrections on Dewey facts by Larry Hickman, who has long encouraged my adaptations of pragmatism, Jane Bunker, editor in chief of State University of New York Press, for her interest and help in this publication, and especially the many participants in the technoscience research seminar at Stony Brook, who have, over the years, read and responded to works in progress. The index was prepared by Frances Bottenberg, also a seminar member.

Chapter 1

What Is Postphenomenology?

This book provides, in four chapters, a perspective on a very contemporary development stemming from my background in phenomenology and hermeneutics as directed toward science and technology. I have coined a special terminology, reflected in the title, *postphenomenology and technoscience*. And while a *post*phenomenology clearly owes its roots to phenomenology, it is a deliberate adaptation or change in phenomenology that reflects historical changes in the twenty-first century. And, in parallel fashion, *techno*science also reflects historical changes that respond to contemporary science and technology studies. It is my deep conviction that the twentieth century marked radical changes with respect to philosophies, the sciences, and technologies. And this is clearly the case regarding the *interpretations* of these three phenomena. I illustrate this by referring to what has been called, in Anglophone countries, the "science wars." The American version, some would hold, began with the 1996 publication of the article "Transgressing the Boundaries: Towards a Transformative Hermeneutics of Quantum Gravity" in *Social Text*. The author, Alan Sokal, was a relatively unknown physicist, and the article was a deliberate hoax designed to show the ignorance of literary theorists and humanities academics. *Social Text* is a radical literary theory journal, and its board of editors was fooled into accepting and publishing the spoof. All of this escalated within the academy, in newspapers, and on the Internet. Stated broadly, the "wars" were about whether or not science is a universally valid, privileged mode of knowledge, culture and value free—this was the stance of the "science warriors." The literati, who were the brunt of the hoax and attack, were thought to have attacked science; they were claimed to be relativists, denying universal and absolute truth, for whom all modes of knowledge are simply subjective (the usual targets here were deconstructionists, feminists, and "social constructionists"). The vast popular discussion, of course, made Sokal "rich and famous," and the aftermath included a whole series of books, articles, and television shows.

This was the American version, but as I had already pointed out in my *Technology and the Lifeworld* (1990), a British version had predated this set of battles. In 1987, *Nature*—surely one of the top science magazines—had included an opinion piece, again by two physicists, T. Theocharis and M. Psimopoulos, "Where Science Has Gone Wrong." Their thesis was that the decrease in support and funding, particularly in the Thatcher era, was due to the relativism of *philosophers of science,* and printed mug shots of Paul Feyerabend, Imre Lakotos, Karl Popper, and Thomas Kuhn headed the article. The objection was that these philosophers of science had undermined the belief in the universality, absoluteness, and value-and-culture free knowledge produced by science. And while this debate did not become as popularized as the later Sokal affair, it did continue on the pages of *Nature* for more than a year, until it was cut off by the editors.

In both of these cases, the "war" was over whether science is to be understood as acultural, ahistorical, universal, and absolute in its knowledge, or whether it is embedded in human history and culture and inclusive of the usual human fallibilities of other practices. Permit me now to *reframe* these incidents differently: One can also see these "wars" as *wars of interpretation.* That is, the context in which these events and controversies take shape includes such questions as: *What* is the most adequate intepretation of science? *Who* has the right to make such interpretations? *From what* perspectives do such interpretations take place? My two examples were of physicists playing the role of science expert interpreters. But what of others? The philosophers, historians, and social scientists? In short, I am suggesting that we "hermeneuticize" these phenomena.

I limit myself to the twentieth- and twenty-first-century context I have set here, roughly the period 1900–2006. With respect to early-twentieth-century interpretations of science, most of the best-known interpreters were philosophers who were trained in or practiced as scientists, including Pierre Duhem, Ernst Mach, and Henri Poincare, in the first decade of the century. These thinkers were trained in mathematics and/or physics. In short, this was a kind of *insider,* or, as it is now known, *internalist* interpretation. Similarly, when historians began to be interested in science interpretation, they were sometimes also trained in the sciences, or they looked at the historiography of science as a kind of heroic biography—great men had great ideas and produced great theories. This kind of history is still favored by many scientists as a preferred history of science.

We can now retrospectively recognize the emergence of both *positivist* and *phenomenological* variants on the philosophies of science. The famous Vienna Circle was formed on the one side, and the Gottingen School, including Husserl, on the other—and recall that Husserl's cognate disciplines remained logic and mathematics! To generalize, virtually all early *internalist* interpreters tended to model their interpretations upon science—and early phenomenology under Husserl conceived of itself as rigorous science. Phenomenology, from its beginnings, was one of the players in the early science interpretation wars.

All of this began to change by mid-century. By the onset of World War II, Husserl had died, and many of the positivists had emigrated to America, where they simply took over most American philosophies of science. Indeed, many emigrant philosophers believed that philosophy itself was equivalent to the philosophy of science. This stance, however, was not unchallenged and I trace its history in briefest form:

- The 1930s through the 1950s remained strongly held by logical positivism or logical empiricism with respect to the philosophy of science. The image of science was that of a sort of "theory-producing machine," which was verified through logical coherence and experiment.
- By the 1950s to the 1960s, a new antipositivist set of the philosophies of science emerged—Thomas Kuhn and kin, those mentioned in the *Nature* controversy—which added both histories and revolutions to the notion of science practice. Antipositivism remained theory centered but added discontinuous phenomena to early logicism. Historical particularity becomes part of interpreted science, "paradigm shifts." This image of science began to be enriched by historical sensitivity. Rather than a linear, cumulative historical trajectory, the antipositivists projected a narrative filled with "paradigm shifts" and punctuated discontinuities.
- The 1970s saw the emergence of new sociologies of scientific knowledge—"social constructionism" and "actor network theory" examined science in its social, political, and constructive dimensions. Science is seen as a particular social practice. Its results were viewed as negotiated and constructed.
- In the 1980s, new philosophies of technology (post-Heideggerian, post-Ellul, post-Marxian) introduced the recognition that science itself is also *technologically embodied*. Without instruments and laboratories, there was no science.

- In the late 1980s and 1990s, feminist philosophies began to locate patriarchal biases in science practice, which in some cases led to new understandings of reproductive strategies in evolution. Science was seen as frequently gendered in cultural practice.

The combined result, decried by the reactions of the science warriors, was that science was now seen as fully acculturated, historical, contingent, fallible, and social, and whatever its results, its knowledge is *produced* out of practices. I contend that by the end of the twentieth century, even those belonging to the analytic versions of the philosophy of science could be seen to have made concessions. For example, Ernan McMullen of the dominantly analytic philosophy of science department at Notre Dame University edited a book called *The Social Dimensions of Science* (1992), clearly acknowledging the now-richer image of science than that of a "theory-producing machine." And Larry Laudan, in his *Science and Relativism* (1990), which is a debate among varieties of analytic philosophies of science, proclaims that all are now "fallibilists."

I take it that this was the consensus at the end of the twentieth century, brought about by a now-widened, more diverse set of interpreters. However, the now enlarged field of interpreters also may be seen retrospectively as a response to the obvious massive changes to both science and technology in the twentieth century. For instance, from 1900 to 2006, one can see that big science, corporate science, and global science are the order of the day. From the Manhattan Project to the Human Genome Project, from physics to biology, there is big science. And the same radical change in technologies should be even more obvious: in 1900, there were no airplanes, no nuclear energy, no computers or Internet, and so on, whereas today these constitute the texture of our very lifeworld. And now my special move: I want to place philosophy, particularly phenomenology, precisely into this scene and interpret it, judge it, through a series of changed interpretations parallel to those used to interpret science and technology. What is philosophy, phenomenology, from a contemporary perspective? Philosophy, too, I hold, changes, or must change with its historical context. This is what produces my attempt to modify classical phenomenology into a contemporary *postphenomenology*. So it is now time to briefly look at phenomenological philosophy, roughly in the same 1900–2006 period relevant here. I do this by first looking at the interrelationship between phenomenology and pragmatism.

First Step: Pragmatism and Phenomenology:

Phenomenology in Europe and pragmatism in America were historically simultaneously born. Both were new, radical philosophies that placed *experience* in a central role for analysis. Pragmatism was first called so by William James (1898), who credited it to Charles Sanders Peirce; William James also was an early major influence on Husserl, but pragmatism was brought to prominence primarily by John Dewey. Note that Dewey and Husserl were both born in 1859, and although Dewey lived longer than Husserl, their philosophical developments were chronologically parallel. But also note that their birth year was also the same as the publication of Darwin's *Origins of Species.* Or, since 2005 was the centennial of Einstein's golden year, 1905, if we also look at Dewey in 1905, we find him at Columbia University, already famous in the philosophy of education after founding his earlier experimental or laboratory school at the University of Chicago. And, if we look at Husserl in 1905, we find him giving his internal time lectures.

In terms of the historical philosophical context at the turn of the century, there were some similarities but also nuanced differences between the pragmatists and Husserl's phenomenology. This can be subtly illustrated in the term *pragmatism* itself. Dewey, in his "The Development of American Pragmatism," says, "The term "pragmatic," contrary to the opinion of those who regard pragmatism as an exclusively American conception, was suggested to [Peirce] by the study of Kant . . . in the *Metaphysics of Morals* Kant established a distinction between *pragmatic* and the *practical.* [Practical] applies to the moral laws which Kant regards as *a priori* . . . whereas [pragmatic] applies to the rules of art and technique which are *based on experience and are applicable to experience* (emphasis added).[1] Now, as we know, Descartes and Kant also play major roles in Husserl's development of phenomenology—but the roles they play are those of an *epistemological* Descartes and Kant, whereas it is the *moral* but also a "praxical" Kant who is used by Peirce! The pragmatic emphasis is on *practice*, not *representation.* This move to praxis and away from representation later repeats itself in virtually all the late-twentieth-century styles of science interpretation.

This different take on Kant is subtle and nuanced, but I want to make a very bold extrapolation from this difference: By using the epistemological Descartes and Kant, Husserl necessarily had to also use the vocabulary of early modern "subject/object," "internal/external," "body/mind," as well as "ego," "consciousness," and the like. And while it is clear that his attempt was to *invert* these usages through the use of his various *reductions*, this vocabulary remained embedded in early phenomenology. This attempt to overcome early modern epistemology, while using its terminology, I contend, doomed classical phenomenology to be understood and interpreted as a "subjective" style of philosophy. The pragmatists, by beginning with the vocabulary of practices instead of representations, avoided this problem. Listen to a contemporary pragmatist echoing this idea: Richard Rorty says, "The pragmatists tell us it is the vocabulary of practice rather than theory, of action rather than contemplation, in which one can say something about truth. . . . My first characterization of pragmatism is that it is simply anti-essentialism applied to notions like "truth," "knowledge," "language," "morality," and similar objects of philosophical theorizing. . . . So, pragmatists see the Platonic tradition as having outlived its usefulness. This does not mean that they have a new, non-Platonic set of answers to Platonic questions to offer, but rather they do not think we should ask those questions anymore."[2]

Returning to Dewey, his early writings contain many essays on the new science, *psychology*. This psychology—although for Dewey the outdated philosopher to be transcended was more Locke than Descartes—proposed to analyze *consciousness*. And whereas Husserl, too, had a problem with psychologism, Dewey again seems to cut to the core more quickly. For him, "consciousness" in psychology is an *abstraction*, whereas experience is broader and necessarily related to other dimensions "if the individual of whom psychology treats be, after all, a social individual, any absolute setting off and apart of a sphere of consciousness as, even for scientific purposes, self-sufficient, is *condemned in advance*. (emphasis added).[3] While Husserl's inversion of Descartes includes "all subjectivity is intersubjectivity," Husserl arrives late at such a recognition. I cannot go much farther here, but one clue to pragmatism's quicker take on the problems of early modern epistemology also may lie in its recognition that there is a biological, evolutionary dimension to "psychology." Put simply, Dewey's frequent model or metaphor for his version of transformational practice is that of an organism/environment model rather than a subject/object model. Again, turning to Dewey's early writings, "In the orthodox view, experience is regarded primarily as a knowledge-affair [Locke/Descartes]. But to eyes not looking through ancient spectacles, it assuredly appears as an affair of the intercourse of a living being with its physical and social environment."[4]

This living being/environment model, for Dewey, is also "experimental," and thus less past or present directed than future directed. Experience in its vital form is experimental, an effort to change the given; it is characterized by projection, by reaching forward into the unknown; connection with a future is its salient trait."[5] (Interestingly, this future emphasis seems closer to Heidegger than to Husserl.) Dewey sees this model as "biological" in some sense, and he imputes this both to one phase of William James's version of psychology, but also to Darwin, whose notion of change-through-time also outlines the points just made. Once again, my contention is that this version of experience short-circuits the "subject/object" detour derived from Descartes—or, in Dewey's case, Locke—and points much more directly to something like a *lifeworld analysis.*

Now, admittedly, I have here the advantage of retrospective vision; I am looking at Dewey and Husserl, pragmatism and phenomenology, from a full century later perspective. But it remains the case that there were resources then contemporarily available from pragmatism, which had Husserl used them would have yielded a *nonsubjectivistic and interrelational phenomenology* along the lines I am now calling *postphenomenology.* This is why I have here paralleled Husserl and Dewey, who were exact contemporaries. This grafting of pragmatism to phenomenology constitutes a first step in a postphenomenological trajectory.

Second Step: Phenomenology and Pragmatism

In my first step, I suggested that the deconstruction of early modern epistemology made in pragmatism could have enriched the beginnings of phenomenology by avoiding the problems of subjectivism and idealism with which early phenomenology was cast. My second step reverses the process, and I now suggest that phenomenology historically developed a style of rigorous analysis of experience that was potentially *experimental* and thus relevant to pragmatism. Dewey's emphasis on his experience-based philosophy was "experimental," or sometimes called "instrumental," but I contend that Husserl's phenomenology contained methods that, had these been adapted in pragmatism, would have enriched its analysis of the experimental. In this case, however, rather than return to Husserlian observations from his texts, I shall instead take these for granted and draw three elements from phenomenology to show how such a rigorous analysis of the experiential takes shape. These include: *variational theory, embodiment,* and the notion of *lifeworld.* Phenomenologists will recognize that all three may be found in Husserl, although I would claim that embodiment was later highly enriched by Merleau-Ponty, and that what could be called the cultural-historical dimensions of the lifeworld were correspondingly enriched by Heidegger. Each of these notions derives from classical phenomenology, but each now takes their shape and role in a contemporary postphenomenology.

I begin with *variational theory:* In Husserl's earlier use, variations (originally derived from mathematical variational theory) were needed to determine *essential structures,* or "essences." Variations could be used to determine what was variant and what invariant. I also have found this technique invaluable in any phenomenological analysis—but as I used this technique, I discovered something other than Husserlian "essences" as results. What emerged or "showed itself" was the complicated structure of *multistability.* My first systematic demonstration of this phenomenon occurred in *Experimental Phenomenology* (1977). Using so-called visual illusions, I tried to show how the phenomenological notion of variation yielded both deeper and more rigorous analyses of such illusions than mere empirical or psychological methods. To demonstrate this analysis, I draw from three example sets from those studies:

In the first example, stage/pyramid/robot, this configuration, an abstract drawing, can be seen as a stage setting. The plane surface at the bottom of the drawing is the stage, while the other surfaces are the backdrops. Thus an apparent three-dimensionality appears—but it also implies a perspective from which this three-dimensionality takes its shape. The POV, or "point of view," is a sort of balcony position from which the viewer looks slightly downward at the stage setting. Here already, then, *embodiment,* or perspectival perception, is implied. But this is only one variation—the *same* configuration could be seen quite differently. Perhaps it is a Mayan pyramid in Central America! In this case the plane surfaces change appearances: the center, upper surface is

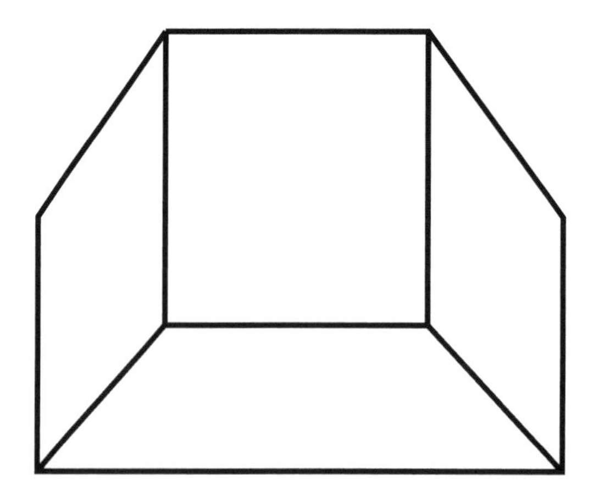

Figure 1.1. Multistable Figure A

now the platform on the top of the pyramid, and the other surfaces are the downward sloping sides. In this appearance, the three-dimensionality is radically reconfigured but remains three-dimensional. And the POV, or perspective, also remains implied—as if we are in a helicopter viewing the pyramid from above. Note too that these two appearances are discrete and different—they are *alternations*, which cannot be combined; they are distinct variations. As an aside to empirical studies, such three-dimensional reversals are well known in psychology—particularly in gestalt psychology, there called "gestalt switches." And while historically the early gestaltists were in fact students of Husserl, we have not left "psychology" quite yet. Now, the first phenomenologically deeper move: I suggest that there is another possible stability here. My story is that this configuration also may be seen as a "headless robot." In this case what was previously the platform of the pyramid now becomes the robot's body. The bottom line is the earth on which the robot is walking, and the other lines are its arms and legs, and—because it has no head—it uses crutches to navigate! In this configuration, three-dimensionality is lost, and the figure is simply two-dimensional. But take careful note: in the two-dimensional appearance, the implied POV, or embodied position, *also changes. Now it is directly before the robot, who is advancing toward the viewer!* This is all fully phenomenological: variant perceptual profiles, examined through variations/implied perceptual-bodily positions, which correlate to and change with the appearances/and, now, more than occur in mere empirical studies.

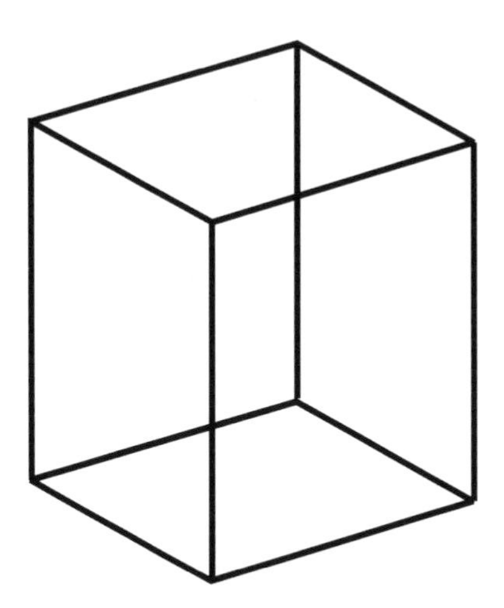

Figure 1.2. Multistable Figure B

My next set of illustrations comes from the famous Necker Cube series. When I was writing my book *Experimental Phenomenology,* I read over 1,000 pages of studies on the Necker Cube phenomenon, all of which recognized the three-dimensional reversal, and a few of which recognized a two-dimensional variant (but usually associated with a "fatigued subject" rather than a noematic possibility within the configuration). Quickly, now, one can easily see that the Necker Cube may be seen as three-dimensional, with a "tilt" switch. Note that there is also a small but detectable switch in the POV, or perspective position, in the switch. To make this into a two-dimensional variation, a new story may be told: I tell you that this is not a cube at all but an insect in a hexagonal hole. The limits of the cube are now the outline of the hole; the central surface is the body of the insect; and the other lines are its legs. Now the figure becomes two-dimensional—and again the POV is directly correlated to the insect. You can easily see that—so far—the Necker Cube has the *same structural set of possibilities as the previous example*, and that the shifts of position, implied embodiment, are all parallel. But since the empirical literature sometimes, though rarely, recognized the two 3-D and one 2-D variations, phenomenology has not yet gotten deeper than gestalt psychology—but it can. Return to the configuration with a new story: what was previously the insect's body now becomes the forward-facing facet of an oddly cut gem. The various surfaces around this central facet are the other facets of this gem—and once you see this, you can immediately tell that this is again three-dimensional in appearance, but in a totally different way than previously as a cube. And, now, if you are learning fast, you can anticipate that a *reversal of this three-dimensionality is also possible.* One is looking from "inside" or from the bottom of the gem and the once-forward facet is not the distant facet. Add quickly, and we have "constituted" *five* variations so far, not three, as in gestalt psychology, and thus once again phenomenological variations go farther than empirical psychology.

My third example set is slightly different than the previous two. In both the stage/pyramid series and the Necker Cube series, the variants were all discrete, distinct, and alternations were not commensurable with each other. Each had multistabilities but discrete stabilities. In this example there is a continuity phenomenon that nevertheless retains its own kind of multistability. This example is the famous Hering Illusion. Here, as one looks at the configuration, the claim is made that the two horizontal lines "appear" to be bent, but in "reality" they are straight. (This appearance/reality distinction, presupposed from modernist metaphysics, is what makes this an "illusion." Phenomenologically, of course,

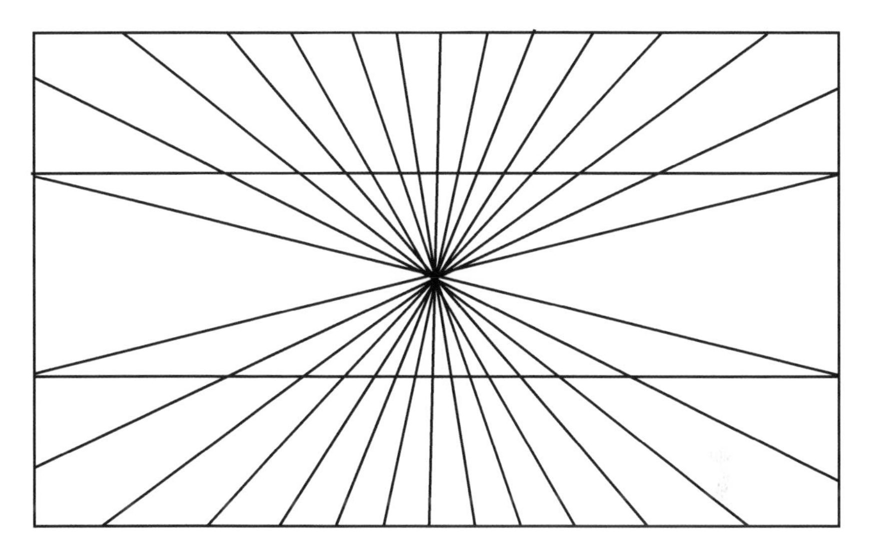

Figure 1.3. Multistable Figure C

reductions eliminate the appearance/reality distinction in favor of what "shows itself.") Now my phenomenological deconstruction of this "illusion" is attainable as follows: focus upon the convergence of lines at the center of the drawing; now "push" this point into distant infinity. As you do this, the horizontal lines *straighten*. Now reverse the process and bring the point that is at infinity back toward your viewing position. You will see the horizontal lines recurve and then straighten out with the three-dimensional reversal. So here again we have multistability, and, as in the other cases, it is related to variations upon two- and three-dimensionality—but also to the context in which straight and curved show a continuous structure. Empirical psychology simply assumed a sedimented and nondepth view, which through deliberate variation shows change. Phenomenologically, perception is not passive but active; holistically, it is bodily interactive with an environment, but while this agrees with both pragmatism and phenomenology, it is the phenomenologically derived *variation* that provides the rigorous demonstration.

We are now partway with step two, the phenomenological enrichment of pragmatism into a postphenomenology. And while I have just made variational theory the central method to give rigor to experiential analysis, the implicit role of embodiment also came into play. Active perceptual engagement, implied in all of the example sets, reveals the situated and perspectival nature of bodily perception—again, an important point repeated by Husserl in his classical analyses (profiles, latent and manifest presences, sometimes even applied to a solid cube in his examples).

At this point I want to make a large leap to an example set now related to *technologies*. While the use of visual "illusions" has the advantage of initial clarity and ease to demonstrate multistability as a phenomenological result of variational analyses, these illustrations also have the disadvantage of being all too simple and all too abstract. This is particularly the case with respect to the weak sense of embodiment in the illusions set. My POV, or perspective, is clear but weak in the sense that I am in a mere "observer" position vis-à-vis the examples. So my next example set will draw from a very ancient, very simple, and very multicultural set of technologies: *archery* (bows and arrows). Although I have researched, and continue to research, the history of archery, I do not believe anyone knows who or where it was first invented. I did meet someone in XiAn who claimed that the Chinese invented archery—in the history of technology, the usual claim is "the Chinese did it first"—but in this case they did not. Some arrowheads date back to at least 20,000 BP; there is an embedded arrowhead in a skeletal pelvis dated 13,000 BP. And, in this case, some European arrowheads date back to 11,000 BP. Then there is Otzi, the freeze-dried mummy found in the Italian Alps in 1991, carbon dated back to 5300 years ago, who had a full archery set with him, two millennia earlier than the 3,000-year-old Chinese treatises on archery. (There is evidence, however, that the Chinese did first invent the *crossbow*, one of which is displayed in XiAn with the chariots recovered there.) In any case, except for Australia, where boomerangs are used, and parts of equatorial jungles, where blowguns are used—rare cultures in which archery never occurred—virtually all ancient cultures had bows and arrows.

My use here, however, is to show how this practice is also *multistable* in precisely its phenomenological sense developed in the earlier examples. Once again I look for variations, embodiment, and now, more fully, lifeworld dimensions. In an abstract sense, all archery is the "same" technology in which a projectile (arrow) is propelled by the tensile force of a bow and bowstring. But as we shall see, radically different practices fit differently into various contexts:

The first example is the English longbow. One famous battle often referred to in European history is that of the English versus the French at Agincourt. This battle was one not only of nationalities but of technologies—the French preferred the crossbow, the English the longbow. Both were powerful weapons, but while the crossbow was somewhat more powerful, it also was slow compared to the rapid fire capacity of

Figure 1.4. English Longbow

the longbow. At this battle, 6,000 bowmen withstood and prevailed over 30,000 infantry and knights. Consider now the material technology, the bodily technique, and the social practice of the longbowmen: the bow was made of yew, usually about six feet or two meters long. It was held by bowmen in a standing position, and the bow was held out in front in a stable position. The bowstring was pulled back toward the eye of the soldier, with four fingers on it, and released when the aim was proper. Arrows were available either in a quiver or stuck in the ground, and firing was fast.

The second example is mounted archery, used by Mongolian horsemen and in the early medieval invasions of Eastern Europe. The horsemen used archery while mounted on speeding horses. While one could say that mounted archers used the "same" bow and arrow technology as weaponry during the Mongol invasions, in another sense there were radical, alternative aspects to horse-mounted archery. First, the bow was short—rarely more than a meter or a little more—made of composite materials (bone, wood, skin and glue), and deeply *recurved*. The power of

Figure 1.5. Mongolian Horse Bow

the bow was similar to that of the long bow but had less distance-gaining capacity. The bodily technique also was radically different. Used while at a gallop, the archer held the bowstring near his eye and pushed the bow outward for rapid fire. (Although not recurved, American Indians used a similar technique for buffalo hunting.)

The third example is "Artillery archery," what I shall call the ancient Chinese archery that utilized the most powerful of all premodern bows known. The pull needed for these long and partially recurved bows was in the 140-pound range. Here the technique called for a *simultaneous push and pull* to launch the arrow, and a unique use of the thumb, with a thumb ring, was required for the bowstring. (I had learned of this technique before actually visiting XiAn in 2004, but during my visit I was delighted to see the terra-cotta archers precisely positioned for this technique!) So what we see again is another stability in which the actual materiality of the bow, the bodily technique of use, and the cultural-historical role this technology plays as a variant.

I am not claiming here to have exhausted the variations, but these three are enough to show that the phenomenological variations that now include considerations of the materiality of the technologies, the

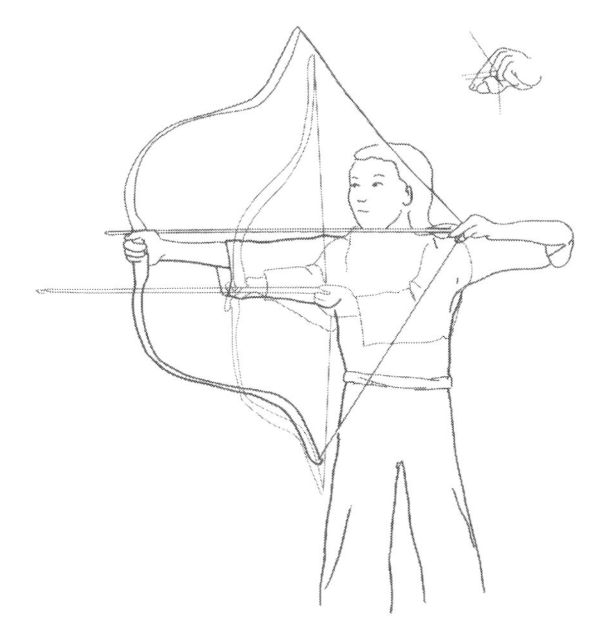

Figure 1.6. Chinese "Artillary" Bow (with Thumb Ring)

bodily techniques of use, and the cultural context of the practice are all taken into account and demonstrate again the importance of variational theory with its outcome in multistability, the role of embodiment, now in trained practice, and the appearance of differently structured lifeworlds relative to historical cultures and environments.

I have now also illustrated the pragmatism to phenomenology and the phenomenology to pragmatism moves needed to outline an initial postphenomenology. The enrichment of pragmatism includes its recognition that "consciousness" is an abstraction, that experience in its deeper and broader sense entails its embeddedness in both the physical or material world and its cultural-social dimensions. Rather than a philosophy of consciousness, pragmatism views experience in a more organism/environment model. The reverse enrichment from phenomenology includes its more rigorous style of analysis that develops variational theory, recognizes the role of embodiment, and situates this in a lifeworld particular to different epochs and locations. There remains one more step to make what I am now calling postphenomenology fully contemporary. That is the inclusion of a "science—better, "technoscience studies" approach to our contemporary lifeworld.

Third Step: "The Empirical Turn"

I began this chapter with a glance at the "science wars" that grew out of issues of interpretation concerning science, technology, and philosophy for purposes here. My contention is that science, technology, and *philosophy* have all undergone major changes through the twentieth into the twenty-first centuries. And while the next chapter will focus on those changes, before concluding my outline of postphenomenology, I turn to one more episode in its construction. In this case we have to move beyond both classical pragmatism and classical phenomenology and into the realm of the philosophy of technology. Neither Dewey nor Husserl made dealings with material technologies thematic. Dewey recognized that psychological experience was a mere abstraction unless it took into account both the physical world and the social world. And while he did parallel Heidegger with respect to the insight that technologies *precede* science, and that science cannot exist without technologies, he did not engage in analyses that would specifically highlight our experience *of* technologies. In Husserl's case, there are few references to technologies at all. The closest he comes—as I have held elsewhere—is in his recognition of measurement practices lying at the base of the origin of geometry, and his recognition that writing raises consciousness to a higher level.

Martin Heidegger was the exception in classical phenomenology, since by wide agreement he may be considered a major thinker at the origins of the late modern philosophy of technology. I also shall return to Heidegger in the next chapter, but in this setting I will "leapfrog" his work in order to outline the third step constituting postphenomenology. That step is what Dutch philosophers of technology have called "the empirical turn," a phrase that has caught on and is now widely used to describe in particular the very contemporary philosophy of technology.

Here is the context: The Netherlands has had a strong tradition in the philosophy of technology, dating going back to the early twentieth century, and one of its main centers today is at the University of Twente. Hans Achterhuis, himself a leading philosopher of technology, collaborated with his colleagues and published in 1992 the book, *De Matt van de Techniek* [The Measure or Metier of Technics]. This book could be thought of as dealing with the early twentieth-century foundations of the philosophy of technology. It dealt with the first twentieth-century founders of the philosophy of technology, including Martin Heidegger and Jacques Ellul, but also Lewis Mumford and Hans Jonas. But in 1997, again with his colleagues, Achterhuis published a second book,

Van Stoommachine tot Cyborg: Denken over techniek in de neiuwe wereld, literally translated as *From Steam Engine to Cyborg: Thinking Technology in the New World*. This book purports to show that a newer generation of philosophers of technology, six chosen from philosophy in America, has shifted the center of gravity by making "an empirical turn." I found this Dutch perspective an interesting one, and thus we had it translated into English (capably done by my colleague, Bob Crease) as *American Philosophy of Technology: The Empirical Turn* (2001).

There are three ways in which Achterhuis sees differences between the classical philosophy of technology and the contemporary philosophy of technology:

- Classical philosophers of technology tended to be concerned with technology overall and not specific technologies. "The classical philosophers of technology occupied themselves more with the historical and *transcendental* conditions that made modern technology possible than with the real changes accompanying the development of a technological culture" (emphasis added).[6]
- Classical philosophers of technology often displayed romantic or nostalgic tastes, thus displaying a dystopian cast to their interpretations of technology. "The issue [now] . . . is to understand this new cultural constellation, rather than to reject it nostalgically in demanding a return to some prior, seemingly more harmonious and idyllic relations assumed to be possible between nature and culture [as in the classical philosophy of technology]."[7]
- Achterhuis notes that the new philosophers of technology took an empirical—or a concrete—turn described thus: "About two decades ago, dissatisfaction with the existing, classical philosophical approach to technology among those who studied new developments in technological culture as well as the design stages of new technologies led to an empirical turn that might roughly be characterized as constructivist. This empirical turn was broader and more diverse than the one that had taken place earlier in the philosophy of science, especially as inspired by the work of Thomas Kuhn, but shared a number of common features with it. First, this new generation of thinkers opened the black box of technological developments. Instead of treating technological artifacts as givens, they analyzed

> their concrete development and formation, a process in which many different actors become implicated. In place of describing technology as autonomous, they brought to light the many social forces that act upon it. Second, just as the earlier, Kuhn-inspired philosophers of science refused to treat "science" as monolithic, but found that it needed to be broken up into many different sciences, each of which needed to be independently analyzed, so the new philosophers of technology found the same had to be done with "technology." Third, just as the earlier philosophers of science found that they had to speak of the co-evolution of science and society, so the new, more empirically oriented philosophers of technology began to speak of the co-evolution of technology and society."[8]

I accept this characterization of the contemporary set of philosophers of technology included in Achterhuis's book. Furthermore, this description is what I am calling the third step toward a postphenomenology. It is the step away from generalizations about *technology uberhaupt* and a step into the examination of *technologies in their particularities.* It is the step away from a high altitude or transcendental perspective and an appreciation of the multidimensionality of technologies as *material cultures* within a *lifeworld.* And it is a step into the style of much "science studies," which deals with case studies.

As Achterhuis correctly recognizes, such a step is not one that occurs in isolation; rather, it reflects precisely the broad front common to most new interpreters of science and technology. The new philosophies of science, the new sociologies of science and feminism, and now the new philosophies of technology all, to some degree, and each in their way, become more concrete in their examinations of what I call *technoscience.*

If this, then, is the contemporary philosophy of technology, then I want to make one final observation about this position compared to both the classical beginnings of pragmatism and phenomenology. As noted earlier, neither Dewey nor Husserl made technologies as such thematic to their philosophies. In Dewey's case, there remained a broad, modernist concern with the natural world and the social world. The experiencer—the human—related to both the physical and the social was thought of as an organism within an environment, in Husserl's case, the "World," or his equivalent of an environment, was also made up of things and of the problematic presence of others, as in the *Cartesian Meditations* and, later, with the historical-cultural-"praxical" world of the *Crisis.* In neither were relations with technologies as such made thematic or specific. With the arrival of the philosophy of technology, which in its dominant form arose from the *praxis traditions* of philosophy—pragmatism, phenomenology, Marxism—the thematization of human experience in relation to technologies produced a changed philosophical landscape.

Such a thematization, however, includes perhaps the farthest-reaching modification to classical phenomenology. In both pragmatism and phenomenology, one can discern what could be called an *interrelational ontology*. By this I mean that the human experiencer is to be found ontologically related to an environment or a world, but the interrelation is such that both are transformed within this relationality. In the Husserlian context, this is, of course, *intentionality*. In the context of his *Ideas*, and *Cartesian Meditations*, this is the famous "consciousness *of* _____," or all consciousness is consciousness of "something." I contend that the inclusion of technologies introduces something quite different into this relationality. Technologies can be the means by which "consciounsess itself" is *mediated*. Technologies may occupy the "of" and not just be some object domain. This theme recurs later in this book.

What Is Postphenomenology?

Postphenomenology is a modified, hybrid phenomenology. On the one side, it recognizes the role of pragmatism in the overcoming of early modern epistemology and metaphysics. It sees in classical pragmatism a way to avoid the problems and misunderstandings of phenomenology as a subjectivist philosophy, sometimes taken as antiscientific, locked into idealism or solipsism. Pragmatism has never been thought of this way, and I regard this as a positive feature. On the other side, it sees in the history of phenomenology a development of a rigorous style of analysis through the use of variational theory, the deeper phenomenological understanding of embodiment and human active bodily perception, and a dynamic understanding of a lifeworld as a fruitful enrichment of pragmatism. And, finally, with the emergence of the philosophy of technology, it finds a way to probe and analyze the role of technologies in social, personal, and cultural life that it undertakes by concrete—empirical—studies of technologies in the plural. This, then, is a minimal outline of what constitutes *postphenomenology*.

Chapter 2

Technoscience and Postphenomenology

In chapter 1, I traced the outline of a modified phenomenology, postphenomenology, as an approach to science and technology studies. In my own case, it was the emergent subdiscipline, the philosophy of technology, that led the way to what today is more frequently called *technoscience*. This term, like postphenomenology, is clearly a hybrid one, combining technology and science. To provide some perspective on the trajectory from the philosophy of technology to technoscience studies, I begin with a brief look at the same basic period followed previously, that is, primarily from the twentieth into the twenty-first century, with only a glance at the very end of the nineteenth century.

I contend here that philosophies of technology, which in the Western context have arisen very recently, are, in effect, philosophical responses to the equally recent changes in technologies. But when does the *philosophy of technology* first occur? If it occurs when it is first titled or *named* as such, then the answer is that it becomes explicit with the publication of *Grundlinien einer Philosophie der Technik,* so termed by Ernst Kapp with a book of that title in 1877. That, at least, is the claim of Carl Mitcham, the premier historian of the philosophy of technology.[1] But, in another sense, naming often comes *late*! For example, the first use of the term *scientist*, occurred in 1854, when William Whewell argued for this term to replace "natural philosophy" which had been the term of use earlier. Yet surely Karl Marx, a contemporary of Kapp's, with his theories of production related to specific technological means, even though he did not title this analysis the "philosophy of technology," must be said to also have done the philosophy of technology. In short, a practice may well precede its being named as such. In either case, it is the end of the nineteenth century that the first glimmer of this philosophical subfield appears.

I do not intend to quibble historically. Rather, my point is that *material technologies* seem to emerge into philosophical interest and awareness in specific form at the end of the nineteenth century—and, as we must realize—in the midst of the historical Industrial Revolution of large and world-impacting technologies associated with this period. Such technologies are so gigantic that even philosophers must attend to them! And, no, I do not want to ignore precedents entirely—much has been made of Plato and his use of craft technologies as examples of skills and knowledges, later captured nostalgically by Heidegger with his glorification of *techne*. Nor should the origins of early modern science in the seventeenth century be overlooked, especially in the oft-quoted claim of Francis Bacon two centuries before Marx and Kapp:

> We should notice the force, effect, and consequences of inventions, which are nowhere more conspicuous than in those three which were unknown to the ancients . . . printing, gun powder, and the compass. For these three have changed the appearance and state of the whole world . . . innumerable changes have been thence derived, so that no empire . . . appears to have excercised a greater power and influence on human affairs than these mechanical discoveries.[2]

Added later to this seventeenth-century list were clocks and the steam engine. And the later, much more Eurocentric version of this list eliminated the historical fact—known to Bacon and acknowledged as such—that, with the exception of the steam engine, all of these inventions were originally Chinese!

Bacon's mind was early modern; he was one of the philosophers of early modern science, one who also recognized early on that science itself was instrumentally and technologically embedded. Still, it remains the case that only with the twentieth century does the philosophy of technology become continuous and thematic. And while the Industrial Revolution produced megatechnologies, and along with these the amplification of human power through machines onto geological scales in mining, railroads, road systems, and factories, with effects so large that technologies could no longer be ignored by the best thinkers. So I now reach that twentieth-twenty-first-century period that is my framework.

In the previous chapter, I noted that the early philosophy of science in the twentieth century began with the "mathematizers": Duhem, Mach, and Poincare. By that scale, serious attention by philosophers to technology began a decade or two later, in the *first generation* of what would become known as the philosophy of technology. Friedrich Dessauer returns in 1927 to Kapp's title, with his own *Philosophie der Technik: Das Problem der Realisierung.* This is the same year as Martin Heidegger's *Sein und Zeit,* in which the famous "tool analysis" takes shape, with hints of the ontological precedence of technology over science. Only two years later, in 1929, John Dewey published in the United States his *Quest for Certainty,* focusing on the role of experimental technology. Karl Jaspers takes account of technology in *Die geistige Situation der Zeit* in 1931; Lewis Mumford writes *Technics and Civilization* in 1934; and Ortega y Gassett does so with "*Meditacion de la Technica*" in 1939. Here, then, is a small indication that many great intellectuals, both in Europe and the United States—in the inter-World War period—turned to the themes of technology, a first generation of the philosophy of technology. The historical backdrop, then even more dramatic with the adaptation of industrial technologies to military uses, had escalated, particularly after the Great War of 1914-1918. As noted previously by Hans Achterhuis, this early philosophy of technology tends to be transcendental, more pessimistic in tone with the Europeans and more optimistic in tone with the Americans—although even the late Mumford also became somewhat pessimistic. Then, after the ravages of the Second World War, thinkers emerged in Europe who set the tone for a much more dominantly pessimistic appraisal of technology—Martin Heidegger, *Die Frage nach dem Technik* (1954); Jacques Ellul, *La Technique* (1954); and a proliferation of critical theorists of the Frankfort School, most prominently, Herbert Marcuse, *One Dimensional Man: Studies in the Ideology of Advanced Industrial Societies* (1964). I cite this sample to illustrate that by the mid-twentieth century, philosophical attention to technology had become full blown, and while not always titled the "philosophy of technology" clearly functions as such.

Before turning more explicitly to the role of phenomenology in this mixture, I want to make a few generalizations about this early movement into the philosophy of technology:

- The philosophically educated will recognize that the philosophers noted all come from what may be called the *praxis* traditions—Marxism and critical theory, phenomenology and existentialism, and pragmatism. Missing are prominent representatives of the positivist and analytic traditions, and there are few connections between this style of the philosophy of technology and the philosophy of science of the time.

- Early work on technologies tended to be transcendental, or "general." The focus was on technology as though it were a single, reified thing. By mid-century, the notion of "autonomous technology" or a runaway technology, over which humans have no control, became a common theme. The literary figures of Faust and Frankenstein were invoked.

- Technology often was cast in a dystopian mode. It was blamed for human alienation from *nature*, it was seen as the cause of the decline of *Kultur*, that is, elite culture, it stimulated the rise of mass man and popular culture, and it pointed to a leveling of all things. In its sociological form, derived from the thought of Max Weber, it also was thought to play a role in the *disenchantment* of nature and the desacralization of the earth.

I identify this set of early-to-mid-twentieth-century philosophy of technology as its first generation. As characterized, this was a more European philosophical phenomenon, although during the same period Dewey and Mumford certainly remained prominent thinkers. North American philosophy of technology did not occur until the late 1970s. Both Edward Ballard and William Barrett published books in 1978 that echoed European themes—*Man and Technology* and *The Illusion of Technique*, respectively. My own *Technics and Praxis: A Philosophy of Technology* (1979) usually is cited as the first North American book specifically called a philosophy of technology.[3]

Phenomenology and the Philosophy of Technology

Before turning to this later-century philosophy of technology, I first look briefly at the specific roles that classical *phenomenological* philosophy played in the first-generation philosophy of technology. This is necessarily a retrospective look. I begin by returning to Husserl: As previously noted, Husserl had little to say explicitly about technologies as such. There are, however, two points at which Husserl formulates notions that can be taken as suggestive for the emergence of the philosophy of technology. One is the recognition that the appearance of writing is a "technology" that allows a new level of meaning development through its inscription process that can be repeatedly read. The second occurs in "The Origins of Geometry," where Husserl almost becomes a *praxis* philosopher by indicating that the idealization and formalization, which eventually becomes geometry, emerges from *measuring practices*. (I shall develop only this example here.) Husserl's contention in the *Crisis*, and followed out in "The Origins of Geometry," is that originary experience, which is characterized as perceptual and praxical, lies at the basis of science. This "origin" is forgotten but can be recovered. He proposes to accomplish this by *inquiring back* in a unique version of "history." It is my own suspicion that much of this is, in fact, patterned upon Heidegger's style of analysis in *Being and Time* (1927), which much of the *Crisis* (1936) seems to reflect:

- ". . . We must . . . inquire back into the original meaning of the handed-down geometry, which continued to be valid with this very same meaning . . . and at the same time was developed further. . . . Our considerations will necessarily lead to the deepest problems of meaning, problems of science and the history of science in general."[4] [This is to be] . . . "an inquiry back into the submerged origins of geometry as they necessarily must have been in their establishing function."[5]

- These origins, Husserl holds, include primary perceptual experience and human practices. "Even if we know almost nothing about the historical surrounding world of the first geometers, this much is certain as an invariant essential structure that was a world of "things," including the human beings themselves as subjects of this world; that all things necessarily had to have a bodily character. . . . Further, it is clear that in the life of practical needs, certain particularizations of shape stood out and that a technical praxis always aimed at the production of particular preferred shapes and the improvement of them according to certain directions of gradualness."[6]

- It is then from the repeated and refined practices that the gradual "perfection" of limited shapes begins to approximate, or point to the idealization of, what become geometrical objects. "We can understand that, out of the praxis of perfecting, "again and again," *limit shapes* emerge toward which the particular series of perfectings tend, as toward invariant and never attainable poles."[7]

There is in all of this almost a Peircean pragmatic point. Peirce claims, "Knowledge is more properly interpreted as habits of acting than as representations of reality, and thus not so much in need of special foundations as being located in historical and social processes."[8] Here, then, one also can recognize in Husserl a certain emphasis on embodiment, the role of perceptual and "praxical" experience, but you may rightly ask, what has this to do with technologies? Granted, these are precisely what *is left out of the Husserlian analysis.*

So I now *add* technologies to the Husserlian analysis by an "empirical" or empirical-historical gloss. The history of geometry, in its traditional Eurocentric version, locates the first geometry and its practices with the Egyptians. My postphenomenological gloss, then, calls for an examination of precisely these technologically embodied practices. I deal with only two practices: first, the measuring practice by which field boundaries are laid out. The annual flood of the Nile, basic to all ancient agriculture in Egypt, would also damage or destroy previously laid out field boundaries. To reestablish such boundaries, surveying instruments were used. The second example relates to the building of monumental architecture, pyramids and temples, for example. Again, technologies or instruments were needed to establish lines, angles, and proportions.

In the case of the Egyptians, the instruments were simple but worked effectively. The basic tool kit related to the *stretching of ropes.* This practice, and related angle establishing equipment, was embedded in a ritual and repeatable practice, as illustrated here:

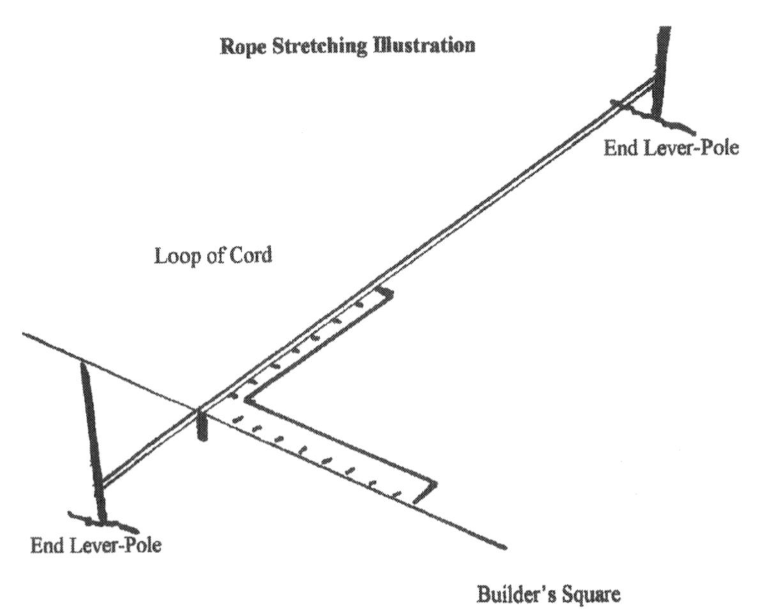

Figure 2.1. Egyptian Rope Stretching

By these means, straight lines for the laying of foundations or the marking of field boundaries are easily grasped. Indeed, when I researched this history, I was amazed, since I too use this practice to lay out straight lines when planting my spring garden—making parallel lines for beans and peas and lettuces. And in my vacation home area, there still remains the town office of the "fence watcher," an official office that cares for boundaries between properties in Vermont. But, even more amazingly, the Egyptian number system uses images of *ropes* for its number system—here are the notations for "1," "10," and "10 squared," all images of ropes laid out. Thus mathematics and instrumental practice show "origins" that belong together.[9]

Finally, the Egyptians were one of at least six ancient civilizations that developed a version of the so-called "Pythagorean Theorum," or the ability to square the hypotenuse that yields perfect squares for whatever purposes are needed.[10] Thus contrary to the Eurocentric master narrative, still believed by Husserl, that geometry is "one" and that once established it develops in a kind of autonomous linear *ideal* history, it appears that geometry had multicultural origins and developed at many times and in many places.

And, wherever and whenever it develops, insofar as it originates from practices, those practices include *technologies of measurement through instruments*, which help constitute a constructed lifeworld shaped by establishing the chosen lines, angles, and shapes with which we remain familiar. Before leaving this example, I hint at three implications that may relate to unstated and possibly unnoted aspects of the history of geometry in their post-Husserlian context. First, once it becomes autonomous, geometrical practice often escapes its "field setting" of making

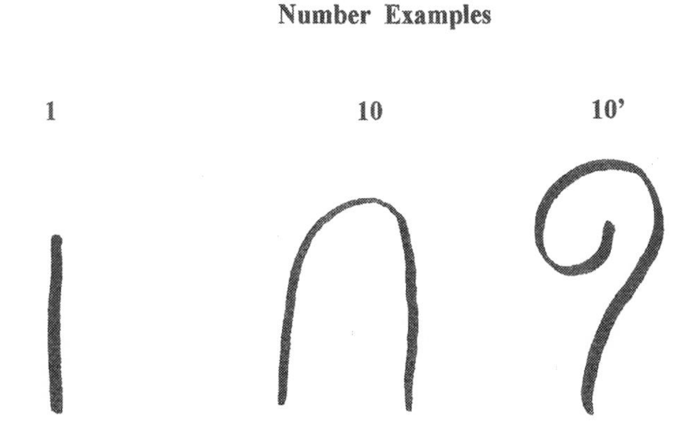

Figure 2.2. Egyptian Rope-Style Numbers

actual measurements as illustrated, and instead it does its "measuring" in a form of writing, that is, through drawings and symbols inscribed on paper, a blackboard, or whatever. Indeed, one material feature of this new practice is that it is *miniaturized* by the use of easily manipulable models, drawings, and the like. It becomes a "modeling" practice in some respect. It remains thus technologically embodied, but it is now placed in a specialized environment, often academic or institutional. It is still praxical, but it now takes place in a different setting. Such practices "inside" of the academy can then be taken "outside" for many different uses. Second, such a practice remains within a *lifeworld,* although it is a differently constructed lifeworld. Geometrical practice reconstitutes cultural perception itself. Even the temple- and pyramid-filled environment of the ancient Egyptians is very different from their own preclassical period of bull rushes and untamed raging rivers. Third, the same applies to science—in a deep sense, science cannot escape the lifeworld, since it too must make all of its measurements *perceivable to embodied humans,* although it may do so through the mediations of measuring technologies or instruments.[11]

From what I have just sketched, it can be seen that phenomenology in its Husserlian form is at best *indirectly* related to any materially sensitive philosophy of technology. This is not the case with Martin Heidegger, however. Heidegger, perhaps more than any other twentieth-century phenomenological philosopher, is directly ancestral to the contemporary philosophy of technology. Within the limits of my context here, I briefly look at only two of Heidegger's lasting contributions to the contemporary philosophy of technology, one from his early period and one from his late period. Both remain positive contributions, in my opinion, although I am quite aware that his role in "deep ecology" and in a widespread influence that appeals to traditional culture and dystopian romanticism remains—regrettably—strong.[12]

The early example arises from his "tool analysis" in *Being and Time* (1927), and it is the most classically phenomenological of his analyses of technology. Interpreters of classical phenomenology often characterize Heidegger's adaptation to be a move to *ontology,* compared to the more *epistemological* thrust of Husserl. That is surely correct in its own way, but I see it as equally a much more explicit turn to *praxis* in a quite pragmatist sense. Beginning the tool analysis, Heidegger says, "The kind of dealing which is closest to us is as we have shown, not a bare perceptual cognition, but rather a kind of concern which manipulates things and puts them to use; and this has its own kind of "knowledge."[13] This is pure praxis philosophy and could have been said by John Dewey! But also note that this shift to the "closeness" of actional manipulation is contrasted to "bare perceptual cognition." I take it that this is an inversion of what for Husserl remained central to his vestigial "Cartesian" context, where perception usually was thought of as perception of things with their qualities, the kind of observing perception common to all early modern epistemology.

The "praxical" "knowledge" that Heidegger attributes to the manipulation of tools, equipment, is not "cognitive" but tacit—and, I would say, *bodily*. Here I briefly trace the key elements of his phenomenology of equipment:

- First, tools are not simply objects that have certain qualities. Rather, what a tool is, is dependent upon a *use context.* ". . . Taken strictly, there 'is' no such thing as an equipment. To the Being of any equipment there always belongs a totality of equipment in which it can be this equipment that it is."[14] This is a rich *field* or *contextual* analysis. In the case of his hammer, a hammer fits into involvements or references to nails, to specific uses, to the task of the project—as well as to a user. "[Equipment] is in order to _____. In the 'in order to' as a structure, there lies an assignment or reference of something to something."[15] Rephrased, *technologies are relative to concrete contexts-in-use.*

- Second, the kind of "knowledge" that arises in praxis is *not* cognitive, Cartesian-type knowledge; it is, rather, a use knowledge in which the tool or equipment becomes a *means* of accomplishment. Phenomenologically, the material tool "withdraws," or becomes "quasitransparent." "The peculiarity of what is proximally ready-to-hand is that in its readiness-to-hand, it must, as it were, withdraw in order to be ready-to-hand quite authentically."[16]

- Yet this withdrawal of the tool-as-object also serves actually to *reveal* or "light up" something else; it implicitly is an account of the environing world or "nature" as Heidegger puts it—or, as I would put it, here is an anticipation of a *lifeworld:* "Any work with which one concerns oneself is ready-to-hand . . . also in the public world . . . with the public world, the environing Nature is discovered and is accessible to everyone. In roads, streets, bridges . . . our concern discovers Nature as having some definite direction. A covered railway platform takes account of bad weather. . . . public lighting takes account of the darkness . . . in a clock, account is taken of some definite constellation of the world system."[17] Technologies, in other words, mediate our way of experiencing a world.

- In what lays the basis for his much later and stronger claims, Heidegger then takes account of what may be called the *breakdown phenomenon*. "When we concern ourselves with something, the entities which are most closely ready-to-hand may be met as something unusable . . . the tool turns out to be damaged or the material unusable. . . . [But] we discover its unusability, however, not by looking at it and establishing its properties, but rather by the circumspection of the dealing in which we use it. When its unusability is thus discovered, equipment becomes conspicuous."[18] This malfunction, or the missing hammer, or the failure of the material, is a breakdown which, in turn, helps us realize that "an equipment" belongs in certain ways to certain contexts. This breakdown phenomenon has become a very paradigm of much post-Heideggerian analysis, down to the present time. (It is used by sociologist Charles Perrow in his famous "normal accidents," by philosopher of science Peter Galison in his analysis of technological breakdowns, and by many others.[19])

The use I wish to make of this is to show how this phenomenon plays an anticipatory role in Heidegger's famous *inversion of the ontological role of the relationship between science and technology*. This is my second example of a lasting influence of Heidegger on the contemporary philosophy of technology. In the early analysis, once a tool malfunctions or breaks, it is an occasion for it to become *conspicuous*. Becoming conspicuous is an occasion for it to be *decontextualized*—at least from its work project. And decontextualized, it may become an *object of examination, present-at-hand,* in short, a "scientific object." In this sense, a scientific examination arises out of, and is dependent upon, a previous, or an *ontologically prior* praxis context.

If *Being and Time* is Heidegger's early and largely implicit development of an understanding of technology, then his later, postwar essays make a philosophy of technology much more explicit. Here, however, I look only at the development of an inverted science-technology relation by Heidegger. The most explicit statement of the inversion occurs in *The Question Concerning Technology* (1977, German 1954), and it is put strongly thus:

- "Chronologically speaking, modern physical science begins in the seventeenth century. In contrast, machine-power technology develops only in the second half of the eighteenth century. But modern technology, which for the chronological reckoning is the latter, is, from the point of view of the essence holding sway in it, historically [ontologically] earlier."[20] Put simply, technology is ontologically prior to science. In 1954, this was a radical claim; today it has a wide number of proponents, including many in science studies as cited in the first lecture.[21] Of course, it should be recognized that "late" Heidegger has come to understand technology in a transcendental, "metaphysical" sense. It is a way of seeing the world, the whole of Nature "calculatively," or technologically.

- This ontological priority of technology over science leads Heidegger to strongly recognize that all modern science is *instrumentally, or technologically embodied.* "It is said that modern technology is something comparably different from all earlier technologies because it is based upon modern physics as an exact science. Meanwhile we have come to understand more clearly that the reverse holds true as well: modern physics, as experimental, is dependent upon technical apparatus and upon progress in the building of apparatus."[22] No instruments, no science.

- Heidegger's ontological inversion reformulates the relations so that science becomes dependent on technology. Science, in a sense, becomes a "tool" of technology: "Modern physics is not experimental physics because it applies apparatus to the questioning of nature. The reverse is true. Because physics, indeed already as pure theory, sets nature up to exhibit itself as a coherence of forces calculable in advance, it orders its experiments precisely [to ask] . . . how nature reports itself when set up this way."[23]

- Finally, the inversion questions the standard view that technologies are "applied" science—this, for Heidegger, is an illusion: ". . . Technology [now transcendentalized] must employ exact physical science. Through its so doing the deceptive illusion arises that modern technology is applied physical science."[24] For Heidegger, technology's ontological priority over science shows that all science must be, in my term, *technoscience.*

In an even shorter glimpse, I introduce one more classical phenomenologist into the precontemporary mix: Maurice Merleau-Ponty. Like Husserl, there is little explicit discussion of technology, but what Merleau-Ponty adds to the antecedents is a very subtle and nuanced discussion of the role of body, perception, and action, that is, *embodiment through technology*, particularly in his *Phenomenology of Perception* (1962, French 1945).

Drawing from the later Husserl and from Heidegger, Merleau-Ponty begins from a notion of embodiment and active perception that, from the outset, is "praxical": "What counts for the orientation of the spectacle is not my body as it in fact is, as a thing in objective space, but as a system of possible actions, a virtual body with its phenomenal 'place' defined by its task and situation. My body is wherever there is something to be done."[25] And again, "My body is geared into the world when my perception presents me with a spectacle as varied and as clearly articulated as possible and when my motor intentions, as they unfold, receive the responses they expect from the world."[26]

But then Merleau-Ponty also recognizes that my active, intentional bodily movement may also *incorporate,* include into its very primary experience, a *technology:* "A woman may, without any calculation, keep a safe distance between the feather in her hat and things which might break it off. She feels where the feather is just as we feel where our hand is. If I am in the habit of driving a car, I enter a narrow opening and see that I can 'get through' without comparing the width of the opening with that of the wings, just as I go through a doorway without checking the width of the doorway against that of my body."[27] While this incorporation of an artifact into bodily experience itself echoes Heidegger's sense of the tool's withdrawal, it becomes in Merleau-Ponty a more subtle and nuanced phenomenon. Finally, there is his analysis of the blind man's cane: "The blind man's stick has ceased to be an object for him and is no longer perceived for itself; its point has become an area of sensitivity, extending the scope and active radius of tough and providing a parallel to sight. In the exploration of things, the length of the stick does not enter expressly as a middle term, as an entity-in-itself; rather, the blind man is aware of it through the position of objects through it. The position of things is immediately given through the extent of the reach which carries him to it, which comprises, besides the arm's reach, the stick's range of action."[28] In short, embodiment or bodily intentionality extends through the artifact into the environing world in a unique technological mediation.

I end here my selective and limited survey of classical phenomenologists and technologies. As indicated, by the mid-twentieth century, many leading philosophers had something to say about technology, including figures such as Ortega y Gassett, Karl Jaspers, and most of the critical theory philosophers associated with the Frankfurt School—but relation to phenomenology often was indirect. There were, of course, important responders to Husserl, Heidegger, and Merleau-Ponty. The responders, writing mostly in the 1960s through the 1970s, include the critical theorists of the Frankfurt School, who were prominent, with Jurgen Habermas, in adapting Husserl's version of a lifeworld; others with positive responses to Heidegger—Herbert Marcuse and Hanah Arendt—and negative ones—Theodor Adorno. But most of the work that was done repeated and echoed the work of the earlier generation. Most retained the notion of technology-in-general, and most produced work that recognized the role of technologies in transforming society and experience by emphasizing the effects of alienation from work and nature and in terms of stifling high culture and producing a "mass man" with a popular culture, one who presumably was replacing the high "Kultur" from which the critics had come.

Thus while recognizing that I am giving short shrift to the admittedly strong impact of, for example, Marcuse's *One Dimensional Man* in the 1960s, I move on to the later explicit rise of the "philosophy of technology" that occurs at the end of the 1970s. I previously mentioned the notion of an "empirical turn," coined by Hans Achterhuis, from the pair of Dutch books detailing the rise of the philosophy of technology. I return to the themes of this book now with a particular concern for the role of phenomenology therein.

Phenomenology and the Achterhuis List

Achterhuis's *American Philosophy of Technology: The Empirical Turn* devotes one chapter each to six contemporary thinkers identified as philosophers of technology; these include Albert Borgmann, Hubert Dreyfus, Andrew Feenberg, Donna Haraway, Don Ihde, and Langdon Winner. Interestingly, four of the six have strong connections to a phenomenological heritage:

- Albert Borgmann, Regents professor at the University of Montana, is the only one of the four educated in Europe with a doctorate from Munich, and he is clearly a neo-Heideggerian. His books reflect many of Heidegger's themes, but he also exemplifies the "emprical turn," especially in his *Holding onto Reality: The Nature of Information at the Turn of the Millennium,* which deals with the rise of information technologies.

- Hubert Dreyfus, at the University of California at Berkeley, with a doctorate from Harvard, has a worldwide reputation for his work on artificial intelligence, the Internet, and computers but is clearly phenomenologically related to the works of Husserl, Heidegger, and Merleau-Ponty.
- Andrew Feenberg, Canada research professor at Simon Frasier University, with a doctorate from the University of California-San Diego, while coming more centrally from critical theory also draws upon phenomenology. His latest book is on Heidegger and Marcuse.
- Don Ihde, Distinguished Professor at Stony Brook University, with a doctorate from Boston University, also draws upon classical phenomenology, although blended with pragmatism and with long-term interests in instrumentation in science, most recently imaging technologies.

One can see, then, that the Achterhuis list is heavily weighted to philosophers who draw from the phenomenological traditions—this is the generation of philosophers prominently publishing from the 1980s to the present. And although my survey has been brief and selective, I now return to the primary theme of this book, *technoscience* and *postphenomenology.*

Technology into Technoscience

Material technologies are *older than modern humans* (*homo sapiens sapiens*)! And while all of our predecessor species have disappeared, we can recognize the antiquity of technologies, material cultures, that go back to very early hominids. Stone Age tool kits go back millions of years, but as Lewis Mumford recognized at least as ancient must have been such technologies as fishing nets, basketry, and the like, which did not survive simply because of the material from which they were made but which must have been part of the "praxical" world of early humans. *Technologies are surely chronologically prior to "science"—at least in any modern sense.* Our oldest ancestors, probably since we acquired upright posture, used technologies to relate to their environments, thus technologies have *always been part of our lifeworlds.*

I offer this bare glimpse at ancient technologies to give a bit of perspective on the radically more saturated technological texture of the late modern into the postmodern world we now inhabit. What my 1900–2006 period displays at its beginning is the major modernist development of industrial technologies—electricity, rail systems, factories, metallurgy, massive uses of hydrocarbons and ores, indeed, what today in America we call "Rust Belt Industries" and, in a sense, the kind of megatechnologies that Heidegger characterized by *Bestand und Gestell.* These industrial technologies have not disappeared, and their march remains, in part, still at work.

But alongside such technologies are those that most people today call the *information technologies*: computers, the Internet, mobile communications, media technologies, and the like. Before I turn to some of the deeper implications of such contemporary technologies for the rise of *technoscience,* I want to take account of some generalizations about "information technologies," particularly as compared to industrial technologies.

- Heidegger called what I am calling industrial technologies "machine technologies," that is, large, powerful, largely mechanical technologies. Steel plants with enormous furnaces, mining technologies with four-story high ore carriers, super-tankers for oil, tree-cutting machines that snip off century-old trees at the trunks—these are all examples of such megatechnologies. I shall risk saying that one *technological trajectory,* as exemplied in these examples, is *gigantism.*

- In contrast, take a set of examples from mid-century of *electronic technologies:* Radios, one of the first such technologies, were at first large, later tiny, while early mainframe computers occupied whole rooms; yet today's laptops equal or exceed mainframe memory and power. Large amplifiers with vacuum tubes were replaced by, most recently, iPods with transistors. My parents' battery-powered, hand-cranked, big box telephone yields to the bare handful-size cell phone that takes pictures, handles e-mail, and reads bar codes for charging merchandise. Here the trajectory is toward *miniaturization and multitasking.*

- Contrast energy consumption in these two reversed trajectories: the electricity-powered aluminum reduction uses 507 billion BTUs to produce 86 billion tons of aluminum, yet when I have my laptop, cell phone, CD player, and LCD video display all on at the same time, these technologies use less electric energy than the hair dryer one uses after a shower. The problem, of course, is that all of these technologies remain in use at the same time, but on different scales and contexts. Finally, both trajectories are *global.* The same chain saw that so vastly multiplies the power of one person to equal an army of stone ax users can be used in Brazil to deforest vast tracts or in Vermont to cut the evening firewood, and the same miniature CD player can play popular or classical music anywhere in the world. But in an admittedly risky generalization, the global decimation of natural resources through mining or deforestation has a different direction taken electronically in the global connection of all peoples and cultures. Satellite television brings here a potential "everywhere."

- One final point concerning the contrast—whereas the factory system of the Industrial Revolution led to the various labor movements, and whereas the boredom of assembly lines often viewed technologies negatively, the contemporary and often miniaturized technologies such as cell phones, Walkmen, and televisions are considered user-friendly, fun, and even empowering.[29]

The first of these technological trajectories took its dominant shape at the end of the nineteenth century, and the second begins in the middle of the twentieth. Some argue, such as had Jaron Lanier, the man who coined the term "virtual reality," that the second trajectory will overcome the first—and it was Lanier who first suggested to me that Heidegger's notion of technology owes more to industrial than to electronic technologies. But there lies hidden in both the secrets of *technoscience,* or of the late modern relationship between science and technology. A word play displays what I have in mind: contemporary science is fully *techno*science, and much contemporary technology is techno*science.* But the interrelationship is not totally symmetrical.

I return to two late Heideggerian observations: ". . . modern physics, as experimental, is dependent upon technical apparatus and upon progress in the building of apparatus."[30] Or, as I put it, contemporary science is materially, *technologically,* embodied. This is one meaning of a *technoscience.* The second claim that Heidegger makes is that the usual classification of such disciplines as engineering is as "applied science"—an *illusion.*" Technology must employ exact physical science. Through its so doing the deceptive illusion arises that modern technology is applied physical science."[31] I agree with both of these claims, but note that historians of science also had beat Heidegger to these conclusions.

L. J. Henderson, as early as 1917, observed that, "Science owes more to the steam engine than the steam engine owes to science."[32] That is because in the early days of experimentation on steam engines, Nicolas Carnot became aware that no matter how efficiently one constructed such an engine, energy was lost. This was the origin of concern with thermodynamics, which in turn led eventually to the discovery of the "laws" of thermodynamics, particularly the "second law." Now, historically, there is something very ironic and interesting about this—that is, science, in effect, derived its understanding of thermodynamics *from observations and experiments upon a technology* rather than from "Nature." It was only later that physicists realized that energy loss also applied to nature, and thus the notion of entropy. Nor is this derivation of a "natural law" from technologies at all unusual. In fact, precisely because early modern science conceived of nature "mechanically," this would inevitably be the case. But it is suggestive of a different perspective on the "ontological priority" of technology over science.

Let us not neglect the other side of the formula; much contemporary technology is also techno*science*. Here the observations are quick and easy—the production of monochromatic, coherent, focused light was not possible until the invention of lasers. But lasers, and the very understanding of coherent light, also were not possible until one could manipulate *photons*, scientific objects, emergent from practices of late modern physics. Yet today the machines that scan grocery store bar codes, the reading of data from CDs to make music, and even my laser pointer all utilize the technologies that this type of science made possible. These are "scientific" technologies now common in our lifeworld. Technoscience is the hybrid output of science and technology, now bound inextricably into a compound unity.

The Contemporary Philosophy of Technology

If, on the one hand, technologies are historically even older than (modern, i.e., H*omo sapiens sapiens*) humans, and on the other hand, contemporary technologies are technoscientific ones, then how can a critical, philosophical investigation proceed? My own answer is—phenomenologically, or, finally—postphenomenologically. I take this opportunity to outline the specific approach I developed over three decades of work. What I have called a *phenomenology of technics* is the focal point of the earlier work on technology. I am gratified that versions of this have been reprinted in virtually every major Anglophone anthology in the philosophy of technology.

The phenomenology of technics is a look at the spectrum and varieties of the human experience of technologies. It took shape first in *Technics and Praxis* (1979) and was reshaped and refined in *Technology and the Lifeworld* (1990). I review here, only in the briefest form, the set or human technology relations developed in the aforementioned book in order to prepare for the empirical studies to follow in the next chapters:

- *Embodiment relations.* Both pragmatism and phenomenology make basic human experience the starting point for analysis. And, I would argue, the later Husserl, Heidegger, and Merleau-Ponty all made practices basic. *Embodiment* is, in practice, the way in which we engage our environment or "world," and while we may not often explicitly attend to it, many of these actions *incorporate the use of artifacts or technologies.* I take it that Heidegger's hammer and Merleau-Ponty's lady's hat feather or blind man's cane are examples of what I call *embodiment relations*, relations that incorporate material technologies or artifacts that we *experience as taken into our very bodily experience.* Such relations directly engage our perceptual abilities—optically our vision is mediated by eyeglasses or contact lenses, our listening is mediated by the mobile phone, or tactilely we feel at a distance the texture of the explored surface *at the end of our probe.* In each of these cases, our sense of "body" is embodied outward, directionally and referentially, and the technology becomes part of our ordinary experience of _______. Moreover, it does not alter our sense of incorporation if the instrument is simple or complex, modern or ancient. In all of these cases, it enters into my bodily, actional, perceptual relationship with my environment. The technology "withdraws," as Heidegger says, it becomes *quasitransparent*, as I say, and thus the technology here is not "object-like." It is a *means* of experience, not an object of experience *in use.* I have formalized this relationship as: (human-technology) → environment. The artifact is symbiotically "taken into" my bodily experience and directed toward an action into or upon the environment.

- *Hermeneutic relations.* I have always held that human-technology (experiential) relations form a continuum. As one moves along a continuum, one finds technologies that engage one's more linguistic, meaning-oriented capacities. Here, while the engagement remains *active,* the process is more analogous to our *reading or interpreting* actions than to our bodily action. There are hints of this in Heidegger's example of the old-style turn signal on old European cars, a pointerlike device that pops up and points as a signifying artifact. Writing, of course, is itself a technology, and it is one of the rare examples partially analyzed by Husserl as a technology that changes one's sense of meaning. But my own earlier examples were drawn from instrument readings. Instrument panels remain "referential," but perceptually they display dials, gauges, or other "readable technologies" into the human-world relationship. And while, referentially, one "reads through" the artifact, bodily-perceptually, it is *what* is read. I formalize this relation as human $\rightarrow$ (technology-world).

- *Alterity relations.* Not all of our relations with technologies are so referential. We may also—again actively—engage technologies themselves as quasi-objects or even quasi-others, hence the term *alterity.* In my earlier work, I used the examples of toys, objects that seem animated and with which one can play. Today, I probably would use *robotic* examples. In Japan I once encountered a robot in a department store who would answer questions about what to find where. Here I relate *to* an artifact—although it is likely that the robot becomes simply an amusing way to be referenced to something other than itself, and thus it reverts to a hermeneutic function. Formalized here, alterity would be human $\rightarrow$ robot (the environment remains background).

- *Background relations.* The robot example already hints at an unattended-to background. As we live and move and engage with an immediate environment, much in the environment is unthematized and taken for granted. And, in any technologically saturated "world" this background includes innumerable technologies to which we most infrequently attend. Once the cold weather begins I turn up my thermostat and once started do not attend to it at all—unless it goes off or breaks down in the Heidegger "breakdown" mode. Once the lights are on, they can be taken for granted until bedtime. Technologies are simply part of our environment.

- *Relational ontology.* As can be seen, in each set of human-technology relations, the model is that of an interrelational ontology. This style of ontology carries with it a number of implications, including the one that there is a coconstitution of humans and their technologies. Technologies transform our experience of the world and our perceptions and interpretations of our world, and we in turn become transformed in this process. Transformations are non-neutral. And it is here that histories and any empirical turn may become *ontologically* important, which will lead us to the pragmatist insight that histories also are important in any philosophical analysis as such.

I have not here developed any of the nuances or finer points of this analysis, but I do want to point out some of the utility of this style of phenomenological analysis: as hinted, the analysis works equally well with ancient as well as contemporary technologies. The skilled scribe in ancient China produces his or her careful style of writing through the "transparency" of the brush, and the better the brush, the better the writing. But the skilled surgeon using laparoscopy is able, similarly, to repair a shoulder rotator cuff with minimal intrusion, again employing his or her bodily skill through the instrument. Here, too, simple versus complex is of minimal experiential difference. Yet, on the other hand, until actually analyzed in each specific case, one cannot simply in advance predict an outcome. Moreover, one must take into account skill acquisition as well. Technologies in use appear differently to beginners compared to skilled users. And as one might expect, multiple outcomes also are likely; technologies tend to be *multistable.* In the next chapter, I turn to a now decade-long research project involving imaging technologies that are of primary importance to contemporary science. This will be an example of an "empirical turn" in my own contemporary postphenomenology, which in turn takes its place among other contemporary philosophies of technology.

Chapter 3

Visualizing the Invisible

Imaging Technologies

An Empirical Turn

Having now schematically outlined *postphenomenology* as a modified or hybrid phenomenology incorporating aspects of pragmatism and turning to the phenomena of technoscience, I now make my own "empirical turn" with a case study such as might occur in science technology studies or STS programs. That is, I examine a particular set of technologies, *imaging technologies*, which turn out to be highly important, particularly in the contemporary sciences.

It is obvious that science and technology, hereafter *technoscience*, has gained enormous prominence in the contemporary world, culturally, physically, and epistemologically. The term *technoscience* deliberately binds two histories, that of technologies that go back as far as all human origins, and that of science, which is usually thought to have a later or *modern* history. Yet however defined, these two histories today belong together in a *hybrid history*. I have chosen to take one strand of recent technoscientific development, imaging technologies, to examine for both its human and epistemological implications. These, I shall hold, are as revolutionary for the contemporary world as any of the changes of knowledge that occurred with early modernity.

I previously argued that *all science, in its production of knowledge, is technologically embodied*. This claim appears in *Technics and Praxis* (1979) and again in a more systematic form in *Technology and the Lifeworld* (1990).[1] This is more than to say that science uses instruments (technologies), but it uses these technologies in unique and critical ways in the production of its knowledge. I am not the first philosopher to have pointed out this importance of instruments—for example, we have already noted that Martin Heidegger noted the necessity for modern physics to use apparatus or instruments. At roughly the same time, Alfred North Whitehead pointed to the same phenomenon with a somewhat different and even stronger claim:

> [Whitehead claims that in modern science] . . . The reason we are on a higher imaginative level is not because we have a finer imagination, but because we have better instruments. In science, the most important thing that has happened in the last forty years is the advance in instrumental design . . . a fresh instrument serves the same purpose as foreign travel; it shows things in unusual combinations. The gain is more than a mere addition; it is a transformation.[2]

This recognition, that science is essentially tied to its technologies, often has been belated, particularly in the philosophy of science. In my narrative, however, I highlight precisely this connection, and in all that follows I trace a number of variables of interaction. First, I take as one such variable *human embodiment*. It is my contention that all science, or technoscience, is produced by humans and either directly or indirectly implies bodily action, perception, and praxis.[3] Second, I also forefront the role of *technologies or instruments* in the roles of scientific knowledge production. Third, I show that since the twentieth century, we have entered what could be called a *second scientific revolution* as radical in its impact upon scientific knowledge as that of early modernity and produced precisely by what can now be called *postmodern* instrumentation embodied in contemporary imaging technologies.

Historical Variations

To situate this second scientific revolution, however, I undertake a brief historical *reframing* of science, looking back as it were, as though good science has always been in some minimal way *technoscience*. This look will be a type of historical variation, dividing periods by the style of technologies used. Simultaneously I take note of practices that correlate with the technologies. When viewed in this way, and commensurate with the observations previously made about the origins of geometry, it will turn out that the history of science can better be understood as a *multicultural phenomenon*. Just as the square of the hypoteneuse was discovered in a number of ancient cultures, so too were other scientific phenomena. I have chosen one primary scientific practice—astronomy—as my exemplar for this history, in part because its history is better known than that of many other sciences, and also because it is of such obvious antiquity. I shall, however, make occasional references to other sciences.

The historical detour I make here varies considerably from the usual Western or Eurocentric master narrative. My own reading of the history of technology and science is one that recognizes that innovations and advancements in knowledge often accompany *periods of strong multicultural interactions.* Part of the reason for this relates to the exchanges of technologies that also occur in such periods. For example, in the standard Eurocentric story, Plato and Democritus usually are credited with the most revolutionary scientific insights. Plato's five geometric ideal cosmic forms (tetrahedron, octohedron, etc.) are cited alongside Democritus's notion of atoms moving in a void.[4] Yet both of these speculations remain just that: speculations. While it is arguable that these speculations were productive—I know of no single scientific development that followed directly out of these speculations—it is not arguable that either had any means of experimentally verifying their speculations. Thus even if regarded as prescient by later generations, neither produced what could be called genuine scientific knowledge. However, later in the highly multicultural era of Hellenistic philosophy, a North African schooled in Athens, Eratosthenes, did produce the first fairly accurate measurement of the size of the earth.

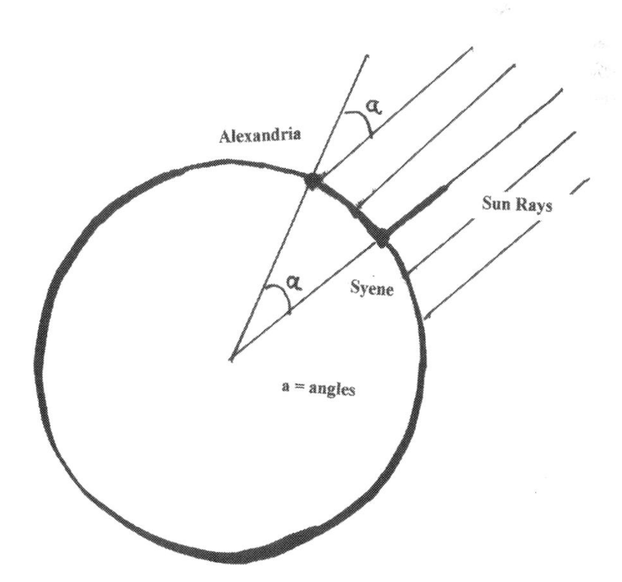

Figure 3.1. Eratosthenes' Earth Measurement

His was an early application of simple geometry with the use of a simple instrument (a gnomon or sundial), which resulted in a moderately accurate measurement.[5] You will note here that I am doing for the Hellenic Greeks just what I did previously for Husserl's origin of geometry—I am looking at historical scientific practices and the ways in which these construct the knowledge of the time.

Similarly, although much later (fourteenth to fifteenth centuries), but directly contributory to the Renaissance revival of science, the multicultural infusion of knowledge from the Jewish and Islamic mathematical and instrumental traditions, along with Christians in the School of Henry the Navigator in Portugal, produced the first accurate navigational maps and measurements that aided the early voyages of discovery emanating from Europe. Here the geometries and instruments were of greater complexity.

In what follows I take astronomy (and associated cosmology) as my science of choice, although what emerges applies to most sciences that today are aided by the latest revolution in imaging technologies. At the heart of this revolution lies a spectrum of new instruments. These instruments, since the late twentieth century, have produced knowledge about the nature of stars, galaxies, and other celestial phenomena that image emissions that exceed the ordinary capacities of human perception itself. Here is how one book, *The New Astronomy,* describes this revolution:

> The "new astronomy" is a phenomenon of the late twentieth century, and it has completely revolutionised our concept of

Figure 3.2. Arab Astrolabe

> the Universe. While traditional astronomy was concerned with [the] study [of] light—optical radiation—from objects in space, the new astronomy encompasses all the radiations emitted by celestial objects: gamma rays, X-rays, ultraviolet, optical, infrared and radio waves. The range of light is surprisingly limited. It includes only radiation with wavelengths 30 percent shorter to 30 percent longer than the wavelength to which our eyes are most sensitive. The new astronomy covers radiation from extremes which have wavelengths less than one thousand-millionth as long . . . to over a hundred million times longer for the longest radio waves. To make an analogy with sound, the traditional astronomy was an effort to understand the symphony of the Universe with ears which could hear only middle C and the two notes immediately adjacent.[6]

While this statement is made with the usual hyperbole that much science rhetoric adopts, it nevertheless points to a new revolution in the sciences. To locate and place this new development, a brief framing narrative regarding a much vaster history of astronomy follows:

- It is probably the case that humans have always been fascinated with the skies, and one practice essential to science, *measuring perceptions,* is evidenced in prehistory. Lunar cycles have been marked upon bones and stones from the

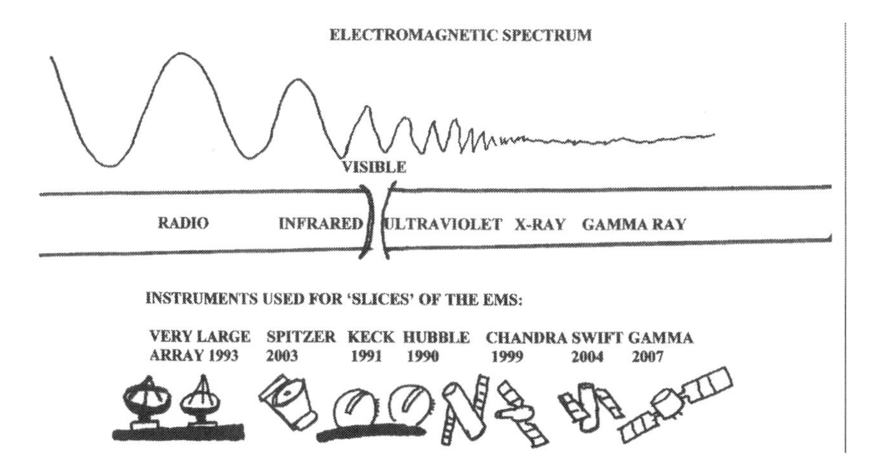

Figure 3.3. Electromagnetic Spectrum

Ice Age (20,000+ BP) (I follow the convention of science dating in which BP means "before present").

- No one knows how old the Austrialian Aboriginal system of color coding the stars (white, blue, yellow, red) may be, nor its use of often barely visible stars in the naming of constellations (10,000+ BP).
- By the time of writing—cuneiform at least (6,000 BP)—vast numbers of tablets from ancient Babylonian astronomy, listing stellar and celestial motions, with classifications, calendars, and solstices marked, can be found by the thousands (hopefully still) in repositories in Baghdad.
- Indeed, the marking of solstices, star motions, and lunar and solar cycles belongs to all ancient cultures: Meso-American, Chinese, Indian, Middle-Eastern, and so on. Many of these cultures produced calendars more accurate than those in Europe until modernity itself.

These ancient practices, often highly sophisticated and systematic, can even be examined for evidence of ancient eclipses, comets, and novae, including, in some cases, predictions of these. I would point out that while strict Eurocentric philosophers of science might point to *demarcation boundaries*, claiming that this is not "true science" because the results are embedded in other cultural practices (such as agricultural, religious, and astrological), the same could be said to belong to Eurocentric science well into early modernity—the embarrassing discovery of horoscopes cast by Kepler a few years ago is evidence that such demarcation, if valid at all, is very recent.[7] Regardless, the ancient knowledge of heavenly cycles and observable phenomena remains valid into the present.

Note too that if we examine this ancient astronomy with the variables I am interested in, that this stage, astronomy, is marked by combining human sensory perception, a measuring, calculative praxis, and the incorporation of some use of technologies, however minimal. Reindeer bone calendars to Inca circular calendars are recording devices, and the use of range-bearing technologies also is ancient. Stonehenge is usually now seen to incorporate an accurate solstice and star motion set of bearings; ancient light shafts from Egyptian pyramids to Hopi light shafts serve similar purposes.[8] Thus highly trained but ordinary human bodily perception and simple technologies belong to the very beginnings of astronomy. And these occur in most ancient cultures, not only those associated with early modern science.

A second step in what I now term *technoscience* produced a first revolution in imaging; this occurs with the invention and use of *optics*, *lens technologies.* In the Eurocentric account, this movement is usually associated with the invention of the telescope in the early seventeenth century, and usually laid at the door of Galileo. Galileo learned of a simple telescope, actually invented by Hans Lippershey in Holland, and he promptly produced a series of improved telescopes and in a series of observations produced claims about phenomena *never seen before but now seen with the telescope's help.*[9] These included (1) the mountains and craters of the moon, (2) the phases of Venus, (3) the satellites of Jupiter, and the observation that began to get him into trouble with the Holy Inquisition, and (4) sunspots.

Without in any way trying to diminish Galileo's accomplishments, they do need to be supplemented with a multicultural perspective: Sunspots, it is now known, were observed and recorded in ancient China as early as 425 BC.[10] Primitive-looking glass in China was first made of dark quartz and may have been the means by which observation of sunspots was made and then recorded. But since there was no dualistic physics of terrestrial/celestial differences, sunspots would not challenge important religious doctrines in China. There is even some claim, however weak, that the naked-eye citing of Jupiter as a "horned planet" by the ancient Dogon in Africa may have been indicators that its satellites were dimly detected perceptually. And, for a scientific account of optics, Al Hazen, the Arabic philosopher, did a treatise on these in 1037, taking into account the *camera obscura*, which later became a model for many subsequent scientific instruments.

Yet wherever they first appeared, lenses introduced possibilities for the construction of radically new scientific knowledge. Lens technologies produced a new, mediated form of human vision. Telescopes mediate human perception in a new way: the embodied observer now takes up a technology that—at first—is literally located between his or her active body and the object observed:

Figure 3.4. Galileo with and without a Telescope

Within this context, vision undergoes a series of phenomenological modifications, with space-time effects. Galileo's moon is magnified—but so is the astronomer's bodily motion when he or she finds it hard to keep the telescope fixed. Thus magnification produces both an *apparent distance* in which the observer and the object observed are "closer" than seemed the case without the telescope, but time, in the sense of a quicker *apparent motion*, also is amplified. The result is that the instrumental mediation yields what was not and could not be seen without the instrument, thus a new, instrument-mediated vision is produced. But we may also note preliminarily that telescopic vision remains thoroughly *analog* vision. Galileo may vary seeing the moon through the telescope with eyeball vision and easily note the "identity" of the moon in two variations.

The magnified, mediated vision suggests the possibility of what I call a *technological trajectory,* that is, a line of development that progressively enhances the discovered capacity of the instrument—in this case, clear magnification. Galileo pushed his own telescope development from Lippershey's, which had only 3X, to 9X, and finally to nearly 30X in his lifetime. Even this improvement was not enough to resolve what Galileo called the "protuberances" of Saturn. Only decades later did Christian Huygens resolve these with an even higher powered telescope that showed the protuberances to be rings.[11]

As radical as lens technologies were for early modern astonomy, they remained *visual analog* technologies. This optical trajectory continued for nearly 400 years and was not exceeded until the twentieth century. What the first revolution accomplished was clearer and greater magnification, but all within the limits, or near limits, of human vision itself. The technological limits remained largely isomorphic with human bodily limits, visual limits.

Having now glimpsed bits of the very long history of astronomy, and in the process taken some note of how that history is differentiated by different human instrumental practices, we can now schematize this history as follows (see Figure 3.5).

One can see that the earliest period did entail instruments, but observations were "eyeball" ones, or unaided. The instruments provided a fixed standard from which to undertake "measuring perceptions." The second period, roughly corresponding to early modern science, used optical inventions—telescopes—to transform vision through magnification, suddenly "increasing" what can be seen and thus increasing a "world." In the next step, new instruments begin to expand observation in ways previously unknown, by exceeding both optical and visible light limits.

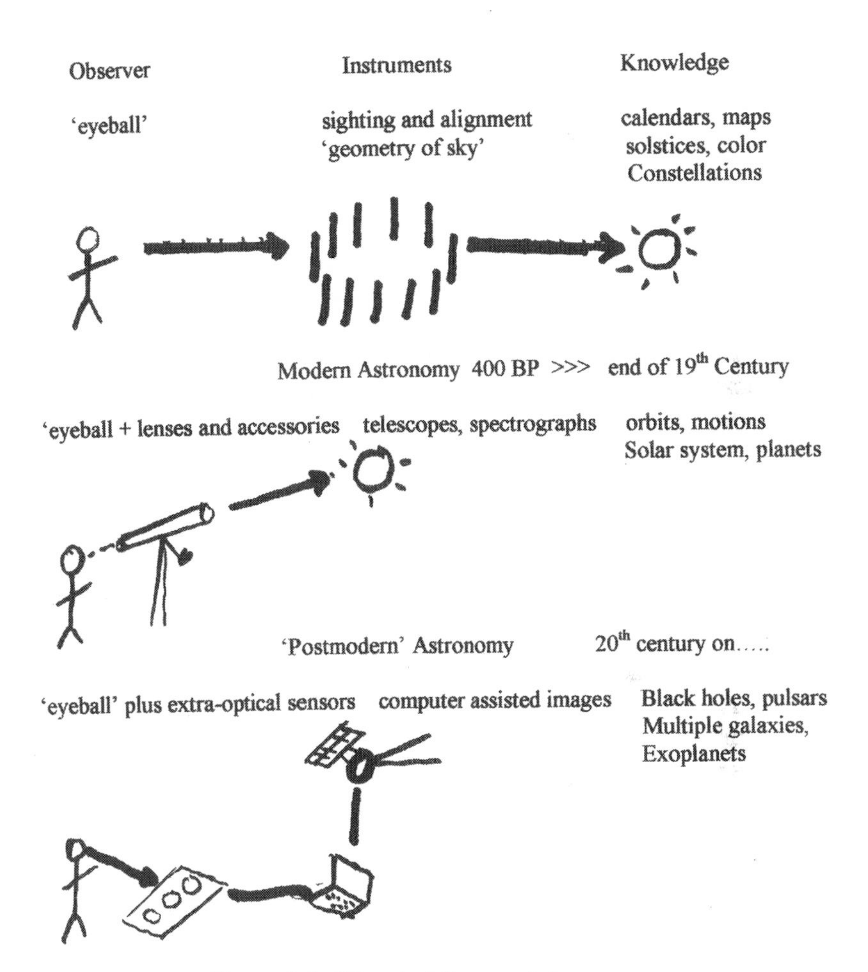

Figure 3.5. Historical Stages of Astronomy

A Second Revolution

I now return to the contemporary period. An interesting question arises concerning how the new astronomy came into being. The usual answer lies in technological developments that took place in the mid-twentieth century. The answer again implies new developing technologies—this time *radio technologies*. At first, if one listened to a radio broadcast during an electrical storm, there was *static*, a hissing in which the noise overcame the broadcast. Acute listeners learned that they could in fact determine the approximate distance of the storm by the magnitude of the static, and its direction, by turning the antenna. It was only later when aimed at the heavens that a fainter version of static was determined to come from outer space itself. Again, citing *The New Astronomy*, "The rapid growth of the new astronomy is due partly to the accidental discovery in the 1930s of radio waves from beyond the Earth."[12] Eventually the "noise" of these extraterrestrial waves was recognized and identified as the background radiation of the universe itself, but not until 1978 was the Nobel Prize given to Robert Woodrow Wilson for this discovery. Later than the 1930s, radio, plus radar techniques, was able to develop into radio astronomy, wherein unseen phenomena could be located through spatially locatable radio sources. Radio astronomy, as an analogue to hearing, brought astronomy to phenomena beyond the limits of optical astronomy. Given that astronomy was so fully linked to vision, a "radio" astronomy must have seemed strange indeed. But it was now clear that celestial emissions somehow *exceeded the limits of human vision*.

Once again, I wish to add a multicultural coda to this development. Within Europe there had already been a nineteenth-century discovery that, while still within what can be called optical frequencies, was a recognition that light radiation exceeded strictly visible limits. William Herschel, experimenting with prisms and early spectographs, noticed that he felt warmth beyond the range of the red end of the visible spectrum; later this felt, but unseen, radiation was identified as *infrared* and could be spectrographically imaged by the late nineteenth century.[13] However, only in the late twentieth to early twenty-first centuries has infrared imaging been able to produce detailed results.

A similar discovery, this time relating to auditory limits, also had been made in ancient China. Chinese scholars noted that there was sound beyond heard sound, experienced by feeling bell vibrations that continued beyond what could be heard with one's ears. They recognized that such vibrations belonged to a continuum extending from hearing, emanating from the bell.[14] Once again, science, scientific discovery, does not belong simply to the Eurocentric master narrative, but my point implies something much stronger—good science may be recognized from and made more robust by being so recognized. In both of these cases, what was recognized was that somehow along a continuum from our sensory experience phenomena exceeded our bodily capacity to detect.

Radio telescopy was the first astronomical imaging to exceed optical limits, but later, other imaging of the *microwave radiation spectrum* became possible. Here we now reach the twentieth century and the imaging revolution proper. First, a few illustrations of precisely this imaging are shown—a series of images taken from different parts of the radiation spectrum.

Each of these images is produced with a complex, but single, set of instruments tuned to some *slice* of the electromagnetic spectrum (hereafter EMS). The first is optical, meaning that it comes from the visible light frequencies of the EMS. The other images, from the X-ray slice, the radio-wave slice, and so on, each imaging only its own slice of the EMS, lie beyond human perceptual range—*none of these emissions can be directly perceived.*

Nor, prior to the twentieth century, did these instruments have the capacity to produce these images and thus the resultant *new scientific knowledge.* Yet from images such as these we now can "see" such phenomena as pulsars, extra-solar planets, gamma-ray pulses, and the vastly expanded set of objects comprising the universe. It should be obvious from such examples that *only through being technologically mediated is the newly produced knowledge possible.*

We should not be blinded here by forgetting how much our very notion of the heavens has changed in this instrumental revolution. Recently I read a remarkable book, *Miss Leavitt's Stars.* It was about the discovery of "standard candles," the bright stars that showed a certain regular variability that could be measured and that gradually were realized to belong to distances previously unmeasurable. I was shocked to learn that as late as 1920, the standard view in astronomy was that there was one, and only one, galaxy—our Milky Way. In today's 14 billion-year-old universe, there are millions of galaxies, classified and arranged into

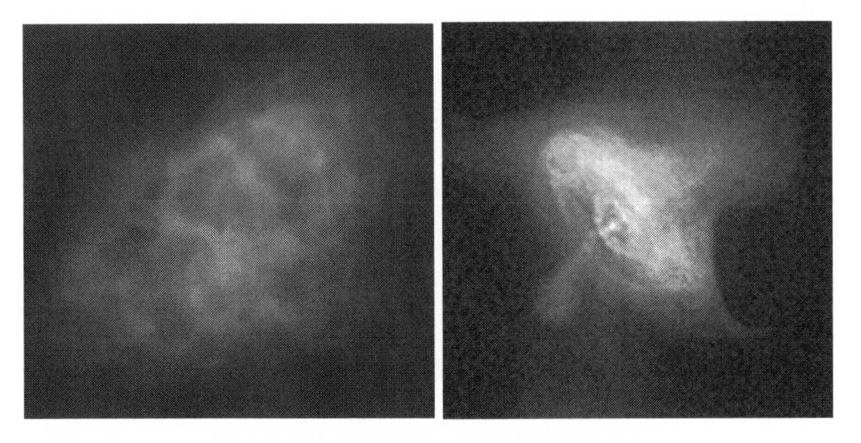

Figure 3.6. "Slices" of the EMS Spectrum

a multitude of shapes (spiral, hourglass, horned, etc.)[15] Along with this now-higher galactic count, the universe also is populated by the array of new phenomena such as pulsars, black holes, X-ray sources, and a variety of astronomical scientific objects (phenomena) simply not known before the twentieth century.

What, now, do these developments show regarding my chosen variables of embodiment, technologies, and technoscience practices? First, as noted, the instruments, technologies, are obviously essential and necessary for the production of the scientific knowledge now emerging from the "new astronomy." If one *reflexively reverses* perspective, then the question of human embodiment can again arise. I argue that we are not now in the realm of the "posthuman," as some have proclaimed. Rather, we now have, with the new imaging, a different kind of human-technology-knowledge relation, a relation that I term *embodied hermeneutic*. There remains a reflexive reference to human embodiment and perception, but it is differently located.

Returning to the image illustrations, in each case the new image produces a *perceivable*, "*readable*" result. The emission patterns, with intensities and shapes, are now *translated by the instrument* into bodily perceivable images, perceived and "read" by the observer-scientist. What I am calling *translation* is a technological transformation of a phenomenon into a readable image. This is one analog to a hermeneutic process, except in this case it is a *material* hermeneutic process, not one limited to textual or linguistic phenomena. Second, because it is perceivable—usually in science's favored visualist modes—it also is available to the gestalt capacities of human vision, which can "see at a glance" the patterns displayed. In this sense, it is a *phenomenological hermeneutic*. So rather than leaving embodiment, the new imaging produces for embodied observers a new way of bringing close something that is both spatially and perceptually "distant."

Many more hermeneutic analogs can be found in the new imaging—I restrict myself to one more. The new imaging brings into play a number of diverse technologies. Light, gamma rays, and X-rays are all gathered by different sensors or reception devices. Each can produce a "slice" of the radiation spectrum. These "slices," somewhat akin to phenomenological variations, can each show some different aspect of the celestial phenomenon. For example, an X-ray slice of the Crab Nebula shows with dramatic results the pulsar source at its center, along with the radiation jets emitted from the rotating center (see previous figure).

Interestingly, if a composite image of the Crab Nebula is shown—this is possible only by computer tomography, which combines the slices into a composite result—the pulsar-jet feature is obscured.

So the first *postmodern capacity,* as I shall call it, of the second revolution is the capacity to image phenomena *not able to be experienced by the body, not perceptible* at all—to direct bodily sensory capacities. But such phenomena are able to be do become experienced if they are technologically, instrumentally *mediated.* Allow me to make the point much more strongly: without instrumental mediation, no experience of such phenomena is possible at all—no instruments, no science.

It might be possible at this point to assume that the new instruments—gamma ray, X-ray, ultraviolet, optical, and radio wave imaging—are simply telescope variants, but they are not. They are compound and complex, and the processing of their resultant imaging incorporates computer and digital technologies. This, indeed, makes them as different from the older optical technologies of early modernity as the optical technologies differed from eyeball observations.

I preliminarily recapitulate what establishes this imaging revolution: First, sensors that can detect emissions beyond the limits of the optical spectrum make possible "extrasensory" imaging. Second, when compounded with digital and computer processing, new capacities of *image construction* become possible. One such process is that of the *conversion of data into image and image into data,* which computer processes make available. Here is a personal example: Some years ago I received an attachment from my older son; I pulled it up and *printed out some 24 pages of gobbledygook:*

My wife, who had a better computer than mine, pulled up the attachment in image mode showing it to be a digital photo of my son, his wife, and a new grandchild!

In astronomy, when a space probe makes a radar image of Venus, its imaging results must first be saved, then converted to data—linear gobbledygook—transmitted to the earth station and reconverted into an image. But there is more to the possible constructability than simply data/image conversions. One also may *manipulate* the image through what may be called simply *enhance and contrast operations.*

I actually do not like the term "false color," since it somehow implies a "true color." I prefer *relative color,* in the sense that one chooses a color or color scheme *relative to the purposes needed.* Relative color is relative to our interests and purposes and yields precisely what we are looking for in a better way. (I have failed to get the scientific community to pay attention to this objection.)

My second example is an image of a crater on Venus, obtained by radar imaging. But here the enhancement is one of increasing the vertical scale—in this case by a factor of 22.5! This enhancement was shown on television in one of those typical hyped science documentaries. Some

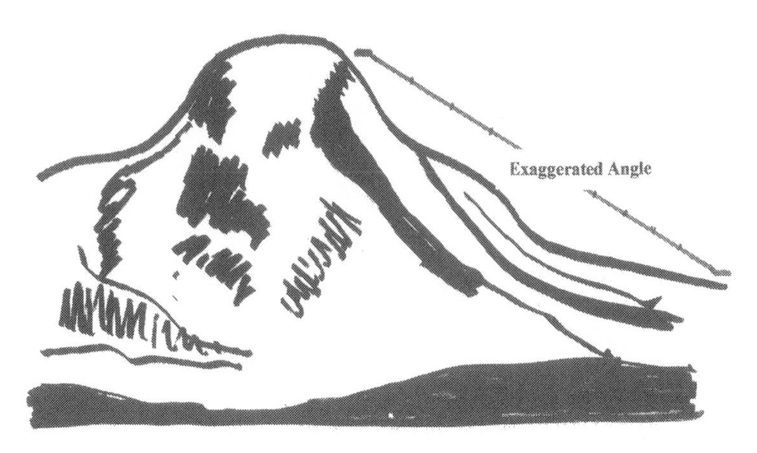

Figure 3.7. Digital Exaggeration: Venus Volcano

astronomers objected, noting that there was no known crater on Venus that exceeded a 3 percent grade, and thus they suggested that there ought to be a "Flat Venus Society" to oppose such an exaggeration![16] Interestingly, I have seen this exaggerated image reprinted many times, including in *The New Astronomy* book I have been quoting—although in this case the subtitle does include a note indicating that the vertical scale is exaggerated by 22.5.

To this point my narrative has been both very narrow—looking only at astronomy—and limited, noting only a small number of processes. However, my primary point is to show how imaging technologies as new developments in technoscience illustrate a revolution for the production of scientific knowledge. I want now to make only two more points.

First, although I have limited my exposition to astronomy, it is easy to extrapolate precisely the same insights into many other sciences. Just one example: medical imaging. In this case the extra-optical imaging actually began earlier than in astronomy. The discovery of X-rays (1895) allowed the imaging of interiors at the end of the nineteenth century. These were immediately applied to medical processes, and early X-rays were able to locate, for example, bullets inside human bodies for extraction.[17] By the late twentieth century the same "slice and composite" processes noted for astronomy had become standard in medical processes as well. Hi-tech X-rays such as CT (computerized tomography) scans, were supplemented by the late twentieth century with magnetic resonance imaging scans (MRI and now fMRI) scans, and these, in turn, by positron-emission tomography (PET) scans. And just as with

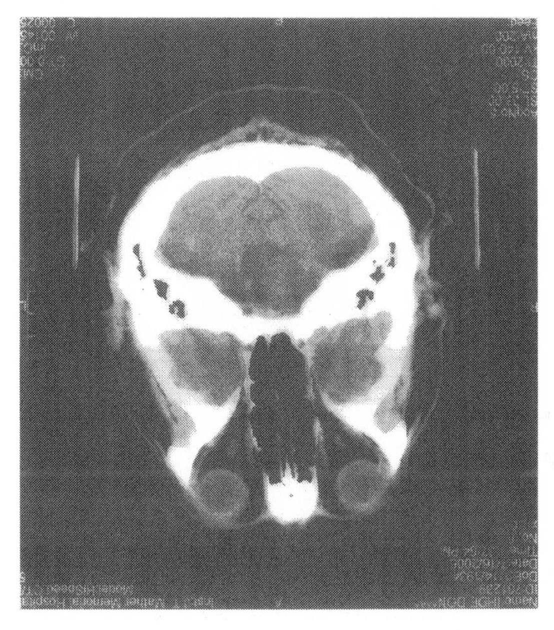

Figure 3.8. Ihde's Brain: CT Scan

astronomy, "slices" and "composites" can and are produced. But whereas in astronomy the "slices" are often the most important, in medicine the composites are often more important. For example, a composite of MRI, CT, and PET scans can show a virtual three-dimensional brain tumor and thus be an aid to the surgeon.

I show this simply to indicate that the imaging technology revolution today has pervaded a vast number of sciences and operates according to the same principles, all of which operate at a level qualitatively different than the previous level of optical scanning and light photography.

Finally, I now return to my simple framing narrative with the features noted concerning the new imaging, with both a forward and retrospective "scan" on these histories to show what and why I make the conclusions I do. For this, I return to my "history" of astronomy chart (see previous figure). If we read this chart forward, it appears that there are long periods of what Thomas Kuhn would have called "normal science," in the sense that there was a growing accumulation of astronomical knowledge, accompanied by an evolution of technologies. Then, in relation to my embodiment, practice, technology variables one can note that ancient astronomy was limited to a vision or perceptual practice that remained constrained by "eyeball" observation. Granted, the ancient astronomer had to have very good eyesight, be rigorously trained, and be patient with respect to repeated viewings of the heavens. For practical purposes, probably from the very beginning of any systematic interest in the heavens, simple instruments or technologies were used within the practices. In addition to recording technologies (calendar bones), other devices probably were used to mark the rising and setting of stars (setting solstices). Thus embodiment and instrumentation stretch back to prehistory. These practices persisted even into the nineteenth century. Caroline Herschel, the sister of William Herschel, was one of the most talented comet spotters of the last century. She used a simple frame to outline the part of the sky she wanted to note, and then with sharp visual practice and perception she was able to identify comets (fuzzy appearance, relatively fast motion) as objects against a recognizable background.[18] And, while within this history there accumulated accurate calendars, knowledge of the types of celestial bodies, solstices, and the like—all valid scientific knowledge by my account—it was also a very long history indeed, at least tens of millennia long.

With modern astronomy, beginning with the seventeenth century, there occurred a technoscientific revolution, one embedded in the invention of lens optics. Here the variables change. There is a qualitative change, in that human vision becomes "eyeball-plus-optics," which transforms the phenomenological space-time of observation through magnification and other optical effects. Now what was previously unseeable with the naked eye becomes visible through mediating lenses. But it remains isomorphic with eyeball vision, to the extent that it is what one would see if one were literally spatially placed at the apparent distance that the lens focus allows. It remains strictly analog in that it is what would be seen if the viewer were literally in the position that the lens phenomenologically yields. Galileo could vary what he saw of the moon through the telescope and the naked eye and easily recognize it was the "same" moon. And once trust in telescopes became sedimented, even when that variation was no longer possible, the "instrumental realism" of lens-mediated vision remained stable. One contemporary philosopher of science, Bas Van Fraasen, still holds that this is the only sort of "realism" that can be claimed.[19] But if this were so, then a great deal of science would simply not be possible. This period, too, was a relatively long one, approaching four centuries, although it is much shorter than the millennia that preceded it. And, as with the ancient period, within it can be detected much refinement and improvement, but within the same qualitative level of practice.

The second revolutuion is again a technologically embedded one into what I might dare call a "postmodern" period of astronomy that takes yet another transformational change of qualitative level. It is the opening—through new technologies—of interrogation and emission detection of radiation from the heavens of nondirectly perceivable phenomena, above and below our perceptual opening to the world. It is no longer analog—except through technologically constructed translations. But I argue that it remains "realist" if by this is meant that a sensor device only operates if it actually detects some emission, but at the same time it is both a constructed and an intervening process that is deliberate and designed. It brings into presence previously unknown phenomena, but it does so by what I have called a *hermeneutic process*, by translating what is detected into images that can be seen and read by embodied observers. The embodiment of observers is thus an invariant in science. This period of astronomy is still no more than a half century old.

Now allow me to reverse the narrative. Instead of looking forward from antiquity to the present, what if we look backward from the present to antiquity? Looking backward, retrospectively, what I draw from this history is that what I have described as "revolutions" are revolutions precisely related to technologies, instruments. In astronomy, these may be thought of as instrumental *embodiments*. Human perception is transformed in each new technological development, situated and placed differently, but implied by each technological context. If we look only at the modern and postmodern scientific revolutions, then it is clear that both are impossible without—and now making a stronger claim about what may have been produced by the technologies associated with science in its concrete histories. In the end, that is why I now call the phenomenon that includes postmodern discovery *technoscience*.

Many enticing implications are associated with this claim. First, as most science or technoscience studies have shown, the older arguments associated with traditional philosophies of science need to be either reformulated or abandoned. One thing is clear: the notion that science progresses by pure theory, which only later can be verified, is one that must be highly questioned and modified, if not abandoned. This has already been done in most contemporary science and technoscience studies.

Second, it is important that our insights need to rely heavily upon "histories," with close attention paid to the practices not only of those that have traditionally fit into a Eurocentric master narrative but those that belong to many cultures and traditions.

Third, we need to develop sensitivity and awareness of the deep materiality of technoscience and take a much more robust account of instruments and technologies as they shape both our lifeworld and our sciences.

Finally, each of the revolutions I have described produced unexpected and dramatic results. There is no reason not to anticipate that we may well invent and discover far more new ways of "seeing" and discovering than we have previously. The richness of the world is far from being known, even in postmodernity.

Postphenomenology Again

This has been an exercise in postphenomenology. It has been "empirical" in the sense of attending to both histories and examinations of actual technologies that have been brought into attention. But it also has been phenomenological in the practices, the role of embodiment, and the application of variations. Many implications follow from such an analysis:

- Embodiment clearly is shown to have different roles and shapes in the history of science. Those differences, as illustrated here, correlate to the different technologies employed.
- Instrumental eras produce different possible sciences. There are "revolutions," but these incorporate "revolutionary" technologies.
- Increasingly, one can see that just as perception, phenomenologically, is active, not passive, the postmodern technologies used by science are active in the sense that they are more and more constructive rather than passive.
- At this stage and level of development, it also can be seen that science *necessarily* must be technologically embodied to produce its knowledge.

Finally, as a postscript, the research project from which this was developed is now a decade-long study of imaging technologies, with special reference to science instrumentation. I have a more detailed program in my *Expanding Hermeneutics: Visualism in Science* (1998).

Chapter 4

Do Things Speak?

Material Hermeneutics

Visual Hermeneutics

Traditionally, *hermeneutics* usually has been associated with linguistic phenomena, particularly texts of various types, with hermeneutics thought of as some set of *interpretive principles.* Indeed, in the nineteenth century, there was an expansion of the notion of hermeneutics such that this mode of critical interpretation was closely associated with the humanities and the social or human sciences. The two thinkers who did most to establish hermeneutics as the method of the human and social sciences were Friedrich Scheiermacher and Wilhelm Dilthey. In brief, both held that *understanding and interpretation* were the unique methods for understanding human psychological and social phenomena. And, in Dilthey's case, this "logocentric" notion of *understanding* was thought to differ essentially from the methods that had proved successful in the natural sciences. The natural sciences, Dilthey argued, proceeded by *explanation* rather than understanding, and explanation was conceived of along hypothetical-deductive lines. This "divide" separated the mode, methods, and outcomes of the human as now opposed to the natural sciences. This "Diltheyan Divide" is often presupposed in much philosophy, even into the contemporary world.

In the twentieth century, this time associated in different ways with phenomenology, three major "hermeneutic" philosophers have expanded hermeneutics even more widely and deeply: Martin Heidegger, Hans-Georg Gadamer, and Paul Ricoeur. In their cases, hermeneutics became a tool of *ontology* itself. But, I contend, in each of these cases, the silent privilege of the linguistic continued to hold sway. It was, as some have termed it, the *continental linguistic turn.* I agree that this privilege to linguisticality, whether in speech, writing, or inscription, or in understanding within a language context, remains the primary characteristic of the humanities and the human or social sciences. It is my argument here that the persistence of the "Divide," and the human science emphasis upon linguisticality, is, in today's context, a mistake! In this chapter I outline what I call a *material hermeneutics* that should belong to both the natural and the human sciences. As I have argued, the natural sciences also are deeply hermeneutical, and, on the other side, the unique hermeneutic techniques developed in the natural sciences have deep implications for the human and social sciences.

This radical claim echoes the theses I previously put forward in *Expanding Hermeneutics: Visualism in Science* (1998). In that book, I, along with many others, recognized that the natural sciences are intensely *visualist*. I claim that while this is historically quite easy to establish, this is a feature of modern *science culture*. This implies, since cultures can differ, that *it could be otherwise*. The previous chapter was a confirmation of this visualist practice. The imaging technologies displayed, whether within the analog range of human experience or transformed and translated into visualizable images, were dominantly "visualizations," which is increasingly the very term used to discuss imaging in most sciences today. Even when the original phenomenon is, for example, auditory, the usual trajectory is to transform the imaging process into a visualization. For example, vocal speech is displayed visually on an oscillograph. My wife uses this type of display in teaching English as a second language in order to improve pronunciation. Sonar, originally an acoustic instrument, yielded to visual display screens. Echocardiograms are displayed as visually graphed oscillations as well (although they may also include an audio presentation as well). It is as if the entire sensorium, for science purposes, is reduced to and transformed into visual form.

Such visualism is sometimes decried as "reductive," as ocularcentric, and therefore a "bias" is alleged. But in the decade of research that I have now undertaken, I gradually began to realize that imaging practices had produced a very sophisticated *visual hermeneutics*. This hermeneutics retains the critical, interpretive work that all hermeneutics requires, but it is more a *perceptual* than a linguistic interpretation. After all, so much of natural science investigates nonspeaking, nonwriting, nonlinguistic phenomena!

Let us return to astronomical imaging, this time taking a different angle upon those practices. I begin with a few seemingly simple geometrical variables: shapes, sizes, distances, and motions. Now we can note from the beginning that cosmology is going to be—at least at first—limited precisely to *visible phenomena*, as previously mentioned, to emissions of visible light. In spite of this limitation, antiquity produced some rather amazing results:

- Eratosthenes' measurement of the earth's circumference was a close approximation to modern measurements.
- The shapes of the moon, earth, and planets were well recognized, and although there were attempts to measure the distances between the earth, the moon, and sun, the distances arrived at were so vast that many measurers doubted the results.
- Ancient thinkers did conclude that the moon was composed of rock, and some surmised a heliocentric solar system. All of this was attained with "eyeball" observations, with simple geometry and instrumentation.

Once optics were used—in early modernity—the awareness of the universe's size began to expand:

- Galileo's telescope revealed the Milky Way to be made up of myriad stars and not a glowing "soup." Jupiter's satellites proved that the earth was not the only body that had such satellites, and the Copernican shift to heliocentrism became irreversible. Galileo also recognized that the planets, through the telescope, changed their disc size, but that all other stars did not—they simply became more intense—and he rightly concluded that this was because they must be much farther away.
- While Giordano Bruno was burned at the stake for his belief that the universe was infinite, the universe had, at the least, vastly expanded in early modernity.
- Yet even in late modernity, while the universe was known to be millions of years old, it was thought to consist, until after Hubble's measurements in the 1920s, of a single galaxy, albeit now light years away from earth.

Astronomy, still limited to the spectrum of visible light, could be said to be "passive" in the sense that whatever could be learned had to be learned from detectable emissions from those distant bodies. Close-up observations, let alone manipulations, were simply not possible. Still, late modernity proved to be ingenious in some unsuspected ways:

- Once again the ingenuity relates to new instruments—in this case the nineteenth-century invention of *spectroscopy.* Newton, in the late seventeenth century, had used a prism to produce the color spectrum, or "rainbow" spectrum. From this device, but for more than a century and a half later, the spectroscope was born. Experimenting with different light sources (the sun, candles, gas torches), and with various apertures (slits, gratings), nineteenth century scientists produced ever-more-distinct spectra.
- The application to astronomy was one that produced classifications of starlight; different stars, particularly those of different colors (red, yellow, blue, and white, as per the Aborigines), produced different spectra.
- Finally, chemical experiments showed invariances between spectra and different chemicals. Sodium, after sprinkling salts into a flame used to illuminate a spectroscope, cast

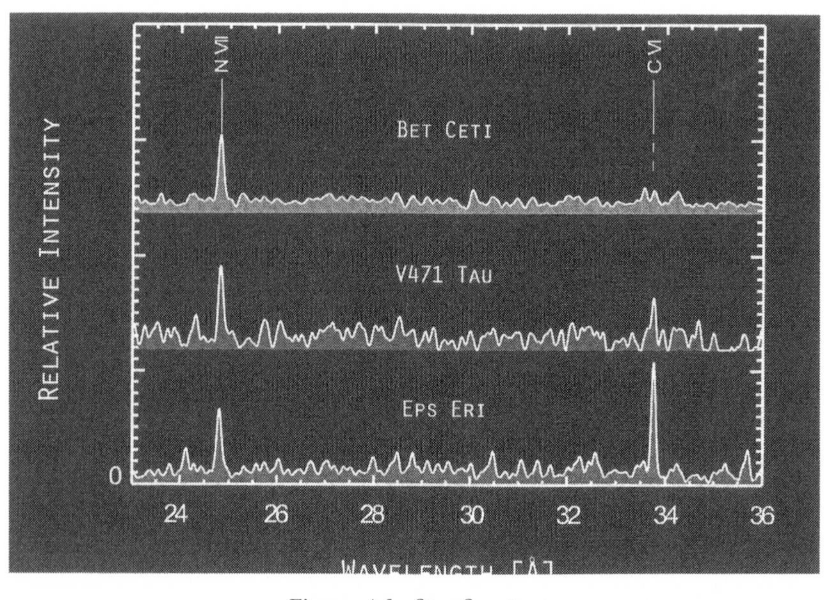

Figure 4.1. Star Spectra

> a distinctive set of lines in the yellow range of the spectrum. This pattern was then recognized to be identical with a distinctive part of the sun's spectrum. Spectra could be used as chemical "signatures" of the composition of the stars.

My point is that instrumental ingenuity, even though limited to the light spectrum, yielded greater and greater knowledge of celestial phenomena. I now make two phenomenological points: First, astronomy, precisely because its objects are so far away, so unreachable, produces a science that fits into an understanding of a "contemplative" observational science that cannot change its objects through observation—this is an antique notion of "objectivity." For the most part, this remains true even of contemporary astronomy. But, second, its observations—science's *perceptions*, if you will, *are not passive.* Rather, they function *very much like phenomenological variations*! In this case, however, these observations are embodied through the use of instruments; they are *instrumental phenomenological variations.* "Eyeball" observations yield to optical ones, but not just the analog ones of optical telescopes, which yield *isomorphic images,* but also *nonisomorphic images,* such as the spectra just noted. Such images must be "read," "decoded," to detect which chemical signature is being presented. Here is already one clue to the implicit hermeneutics embedded in science practice—a phenomenological hermeneutic at that. If one cannot manipulate one's object, then one can manipulate one's instrument.

Not all science is as limited as astronomy is to objects that are so distant that they cannot be "touched," manipulated. The older, classical philosophy of science followed a tradition that effectively retained the notion of a contemplative, observational practice, but with the twentieth century this too began to change. The new sociologies of science, the "strong program," but even more *actor network theory,* which focused upon laboratory life and practices, began to "follow the scientists around" to take note of their practices. Bruno Latour and Steve Woolgar published *Laboratory Life* in 1979—and this was also the year of my *Technics and Praxis.* By taking note of "hands-on" science, embodied in laboratories and instruments, one could begin to understand that late modern science was manipulative and *interventional.* Ian Hacking, coming from a more classical analytic philosophy of science, recognized this as well with his 1983 *Representing and Intervening.*

Finally, before I make my material hermeneutic term, I wish to say good-bye to astronomy one more time:

- For practical and instrumentally limited purposes, early modern astronomy was limited to the solar system—which it equated to the universe. Its revolution was the shift from the Ptolemaic to the Copernican systems, and this "revolution" so impressed philosophers that Kant used it metaphorically as "The Copernican Revolution" for his philosophy, and later Thomas Kuhn used it as a model for his *Structure of Scientific Revolutions.*
- Only in the twentieth century did the notion of the universe expand to its present age of 14 billion years and its content of millions of galaxies with their black holes, pulsars, brown dwarfs, and so on, ad infinitum.
- We also are repeating one pattern of early modern science just now, in that for the first time we are introducing *interventional* technologies into explorations. These more interventional instruments remain limited to the solar system: Cassini, today exploring Saturn; the Mars rovers roam its surface; even a landing and a cannon shot at an asteroid begin to produce interventional science into astronomy. There remains, of course, a very big leap to even the closest star, 4.5 million light years away—think of the patience that would be needed to send a light speed probe taking 4.5 million years, and then awaiting the answer in another 4.5 million years!

Material Hermeneutics

With the natural science examples just used, notice that the object realms investigated usually do not contain "linguistic" dimensions. There are no texts, no speech, no propositional or rhetorical expressions. To observe, whether in the limited passivity of astronomy or the highly interventional practices of particle accelerators, is to enact the questions asked through material, instrumental means. Materiality, in a double sense, pervades the natural sciences, both in the form of what is investigated and in the instrumental modes by which the investigation proceeds.

In contrast, consider the kind of phenomena most typically investigated by humanities and human science disciplines:

- For antiquity, we seek texts, inscriptions, and other forms of *written language;* history is differentiated from prehistory on precisely this basis.

- Or, we seek to hear what participants have to say. In a contemporary world, the media bring us what observers of events have to say, or we hear speeches or read them in the newspapers or on the Internet.
- In philosophy, one looks for arguments; in literature, one looks for rhetorical form or metaphors.

These disciplines are focally oriented to the full range of linguistic phenomena, and they construct their narratives and reconstructions based upon linguistics.

There are, of course, border disciplines, practices that necessarily have to utilize more material investigations—for example, archeology, physical anthropology, and the various newer sociobiological disciplines. And I shall draw from examples from some of these to explore what a material hermeneutics might imply for humanities and the human sciences. First, however, a very simple phenomenological experiment: *How can we hear, or give speech to, that which is silent?* This is a strong metaphor for what I am going to examine.

I have long been interested in acoustic/auditory phenomena and published a book, *Listening and Voice: A Phenomenology of Sound* (1976), republished in an expanded edition, now including new work in acoustic technologies, *Listening and Voice: Phenomenologies of Sound, Second Edition* (2007). So, for my metaphor, I return to discoveries noted therein:

- Within our sensory lifeworld, much of what we *see* shows itself visually to us—but auditorily remains silent, as with a desk. But a desk can be made to "speak" or sound by tapping it. Note, however, that while it is true that I "hear" the desk when tapped; I do not hear the desk in isolation—I hear the desk *plus* the sound made by whatever material thing I use to make the sound. In short, I hear a *duet,* a simultaneous sound of knuckle and desk.
- If I am scientific, I will examine some variations, for instance, by tapping with my pen, then a blackboard eraser. In all of these variations I learn more and more about the desk in its "singing." It is solid; it sounds its materiality; it sounds different than the wall.
- Exploring further, if I tap and give voices to other objects, then some I find to be hollow, some solid; in short, sounds produce *interiors,* which in an analog to vision remain hidden from vision. Such phenomena were of special importance

> to late modern medicine. Auscultation instruments, such as the stethoscope, could be used to detect all sorts of bodily phenomena: heart murmurs, breathing obstructions, congestion in the lungs, and so on.

Now, while I am using these phenomenological and instrumentally phenomenological events metaphorically, to suggest that that which is silent, unheard, can be given voice, I am doing this by introducing a material hermeneutic process that, while already practiced in much science, could transform our humanities and human science practices. I begin with one very dramatic example from recent times: the discovery and analysis of the freeze-dried mummy, Otzi.

Otzi, or the Iceman

On September 19, 1991, two Alpine hikers on the border between Austria and Italy came across a corpse protruding from the glacier. It was obvious that whomever this body belonged to, it was old, indeed, possibly very old. But only its top part was protruding. Several clumsy attempts were made to extricate it, and early guesses were that it might have been a skier or hiker from a few centuries earlier. Soon, however, strange tattoo markings lower on the body appeared, and various gear was found with the corpse—an ax, bow and arrows, bearskin hat, fire-making kit, medicine kit, and other artifacts.

Who was this corpse? And how old was it? (I here divide my brief narrative by early responses and an approximation to the tale that would have been dependent upon pre-twentieth-century archeology-anthopology.) Newspapers began to speculate, stretching the time back five centuries, but the first "expert" analyzing the corpse, site, and artifacts was a prehistorian, Konrad Spindler, who soon wove a tale: "Otzi," as the corpse became known, had probably wandered up into the mountains, got lost in an early snowstorm, and froze to death and died on the spot in mid-fall. Not all of the artifacts were discovered at the same time, but those Spindler had, including bow and arrow, ax, and the kits mentioned, led him to place Otzi—at most—in the Late Bronze Age. That, possibly, would push his age back to 3000 BP. His ax, however, was not bronze but copper, and its design was similar to axes depicted in the statue menhirs of northern Italy dating to a millennia ago. This was an anomaly indeed.

One reason I chose the Otzi example is that associated with this find there are no linguistic phenomena at all—no inscriptions, no texts, no records from the times. To highlight my claims about a material hermeneutics, using science instrumentation to produce knowledge, I began with a suggestion as to what knowledge concerning Otzi could have come from classical archeological analyses. This I now contrast with the subsequent analyses using instrumentation available since 1991 to the present, that is, the postmodern analyses that parallel my earlier imaging case:

- First, Otzi's age—everyone today is familiar with carbon 14 dating, the most accurate dating for phenomena within the scale of time here—and in Otzi's case, that date turned out to be 5300 BP!
- This date, far older than anyone finding Otzi, including Spindler, resolved the ax anomaly. Copper, plentiful in the Italian Alps, precedes the discovery of bronze. Italy's Bronze Age is usually thought to be, at most, 3800 BP—Otzi's ax is a millennium and a half older (5300 BP). That its design persisted in the similar-looking ax heads imaged on the later menhirs is not surprising, since technological design often persisted for millennia in antiquity.
- It is interesting to take note of the bow and arrow set, the fire tool kit, and the medicine kit (which included mushrooms), since these reveal practices that Otzi knew and undertook.
- Once the micro-processes of science technologies come into play, the "real fun" begins. A CT scan revealed that Otzi had an arrowhead under his shoulder blade—speculation concerning the arrowhead underwent two periods. The first concluded that he may have died of this wound, but since projective blades sometimes remain in wound sites without mortality, some argued that this may have been an old wound. Only later, again with finer instrumental analysis, was it possible to see that a major artery had been severed, and thus this was the likely cause of death.
- Paleopathology, employing DNA, mass spectroscopy, and other micro-techniques, led to the unraveling of Spindler's tale. Otzi's stomach and intestinal and abdominal areas contained pollen from the hop hornbeam tree which blooms only in the late spring—Otzi thus died in the late spring. He also had eaten mountain goat, red deer, and a bread made of einkorn wheat, one of the earliest domestic grains in Europe. These discoveries were results of DNA analysis, mass spectroscopy of hair, and electronmicroscopy investigations.

- Otzi also had a moss remnant in his gut, isotopes of that identify mosses that grow only in a nearby area down the mountain; a fingernail showed that he had been seriously ill three times in the preceding year (evidenced in striations on the nail), and again isotopes via mass spectroscopy of his tooth enamel suggested a history of living in two different areas in two different times.
- Otzi's clothing included stiched shoes, layered clothing for cold protection, and kits to deal with fire and health.

Thus from the knowledge now produced, a narrative can be constructed that shows Otzi living in two areas not more than 80 kilometers from his final site, living a life with a mixed diet of meat, bread, and vegetables but also having health problems, including arthritis (shown from bone analysis), and probably employing folk medicine (the tattoos) as hoped-for cures. This yields a life story more detailed than one might expect.

I take this as an example of a material hermeneutic, in which "things" are given voices: pollen, grain, metal, and tooth enamel have all "spoken" in spite of being situated in a context that itself is without proper linguistic phenomena. This is also an application of natural science embodied practices upon an object of human history and the human sciences. The example shows that a fairly rich narrative can be constructed, relevant to the human sciencs, in spite of the absence of linguistic phenomena.

I am quite aware that a material history without linguistic dimensions also can have gaps and problems. For example, once we go farther back in prehistory, even the material culture is spotty. There is a long history of Stone Age tools, such as Acheulean "hand axes":

Figure 4.2. Acheulean Hand Ax

But surely, as Lewis Mumford had surmised, the people of this time also must have had *soft* technologies that did not survive—baskets, nets, and the like. What follows is evidence of this—a "Venus" figurine, from roughly 25,000 BP:

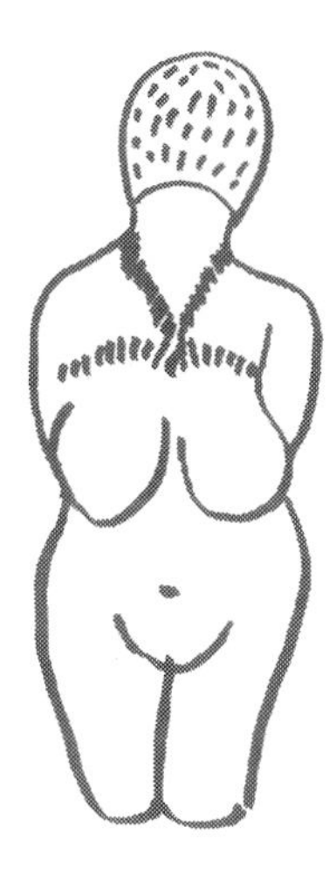

Figure 4.3. A "Venus" Figure with Textiles

Her headpiece—and there are other examples of other stone female figures with associated articles with similar design—is woven or knitted. Textiles thus go back at least this far into prehistory. (Interestingly enough, the first person to note this design feature as a woven pattern was a woman fashion designer turned archeologist, who publicized her findings just a few years ago. No male archeologist seemed to note this!)

If one moves into historic periods, it is possible to parallel textual and linguistic evidence with the material evidence suggested. In a previous work I sketched out two variants of a material hermeneutics of examples including both linguistic and material dimensions.[1] In one, Viking invasions of England, the written accounts portray the Vikings as robbers, plunderers, and arsonists. But the material account, which looks at coinage, burial practices, and the changes in the English language itself, shows that the Vikings also were traders, bringers of a preparlimentary legal system, and settlers who rather quickly assimilated into Saxon culture. The material hermeneutics reveals the written accounts to be partial and in some ways to show phenomena that are in tension with the written accounts. I shall not develop that example here but instead look briefly at the role of materiality, technologies, in often-neglected ways in which humanities and the social sciences themselves are transformed.

Material Mediations

In the previous studies concerning the practices of the natural sciences, I have contended that and shown how instrumental changes affect changes in the sciences using such technologies. New technologies make new science possible. But is the same thing the case in humanities and the social sciences? My answer is "yes," but to show how this is the case, one must attend to the phenomenological interrelation between the technology users and the productive outcomes of the humanistic practices. In my examples in this section, I have to revert to some degree of speculation, since few studies have explored the phenomena I discuss.

My first example relates to musical performance. Clearly, vocal singing can follow a number of styles—an operatic aria can be quite long and intricate; church hymns can have variable verses; and some oral traditions open with a display of the range of the singer before coming to the song itself, which then shows how this range is vocally displayed. In 1877, Thomas Edison invented the first "phonogram," later called the "phonograph." Its early intended use was to record telephone messages, but by the end of the century it began to be used to reproduce musical performances.[2] The reproduction was mechanical, with a diaphragm taking the sound vibrations and translating these into indented patterns on a cylindrical tube covered with tin foil. But there was a severe limitation—the tube could record only two to three minutes of performance. By 1908, several other companies were able to record up to four minutes of performance.

Singers who wished to be recorded and then heard by the public thus had to produce *songs* that were time limited to one of these mediums. Thus began the tradition of the two-to-four-minute song! Here was a singer, adopting to a technology, but also establishing a tradition. Even many iPod songs remain within this time limit tradition. In passing, it is interesting to note that some of the earliest recordings were aimed at "high-end" consumers and were opera arias—but these did not catch on, since most arias were longer than the time constraints of the technology and thus had to be edited![3]

My second example comes from early modern "natural philosophy" and relates to the medium of *letter writing*. The seventeenth and eighteenth centuries were probably the apogee of interphilosopher letter writing. These "Enlightenment" times saw frequent correspondence between major thinkers. Biographers found, in preserved letter correspondence, all sorts of clues and idea exchanges that helped establish the humanities and social sciences as logocentric disciplines. Think of the vast correspondence accumulated by Descartes, the Leibniz-Newton controversy, and then one of my favorite eighteenth-century figures, Hamann, who literally corresponded with virtually every major figure of the day and whose correspondence inspired Søren Kierkegaard. And, while I have no statistics on this, my suspicion is that most letters were several pages long, but that most did not reach article length.

Skip to the electronic era: Where are letters? The handwritten letter as a means of communication between intellectuals, I wager, has almost totally disappeared. Its replacement—electronic communication, e-mail. I am not nostalgic, and I admit that this is the medium I myself most use. But, as with the phonograph, notice the change in page size. It is rare that I ever receive an e-mail longer than two pages; by far, most are a short page or less. This is a quick medium, lending itself to speed and shortness. But have you noticed a compensation? The *attachment*. An attachment is usually an article-length communication, a draft of an article or a lecture, clearly longer than a letter but shorter than a book. (In the sciences, it is the prepublication.) Now, I would argue, this is the contemporary version of the older "Enlightenment" letter.

My point here is simple: the human-technology interaction is one that allows for different trajectories of use, for different possibilities, those that are clearly *non-neutral,* but also short of anything like a deteminism. And the change in technologies produces changes in what and how ideas are communicated, including those of philosophers. Here too is an effect of what I am calling material hermeneutics.

Giving Voices to the Unheard

I now return to my acoustic metaphor concerning "giving voices to things," but I do this by turning to effects produced by acoustic versions of the same contemporary constructive technologies that I have used to illustrate imaging technologies as used by science. My examples here also should show something about another earlier claim I made: *If science is culturally visualist, then the suggestion is that there could be other cultural variants,* and my acoustic examples will precisely suggest that.

My first example is a playful one. Last year I gave a lecture at an art and science exhibit at Ohio State University, and the primary "scientific" art form was holography. My own lecture included some uses of new acoustic technologies that used electronic and computational techniques. I played a number of sound clips, and, as often happens, in the audience was a person involved with precisely artistic uses of acoustic technologies. Daniel Jollieffe presented me with a recent CD of his which, when played, sounds very much like a minimalist composition. The sound was recorded from a digital piano but was produced by no human player. Nor was it composed, although the sound was reminiscent of Philip Glass or Steve Reich. The notes were repetitions, playing within a small note range yet somewhat enchanting.

What the CD does not show, however, is how this music is produced. The title of the piece is "Ground Station," and the music is nonhuman. The piano is a digital electronic piano; in turn, it is hooked up to a computer, which is then also hooked up to a GPS (global positioning satellite), which is hooked up to an antenna on the roof. Here, then, is how the music is produced:

- GPS units locate and "read" the signals of geo-stationary satellites, that is, satellites that stay in the same place relative to earth and its movements. But they are not really fully stable; *they wobble*!
- This wobble is picked up and corrected by the GPS, and the corrections are fed to the computer, which in turn, using an algorithm, feeds into the piano in real time, turning these wobble signals into "music."

At this point we can realize something very interesting, curious, and profound: the computer capacity to convert image to data or data to image *does not have to be visual—it also can be acoustic*! And if the listener knows about this production, she or he can even learn to detect what sort of wobble there is—repeated notes are the same wobble angle, and so on.

My second example is considerably more complex. Dutch physicist Felix Hess has become a scientifically talented "performance artist" whose primary works are often acoustic or partially acoustic. But before playing an example, more initial background may be helpful here. As we noted, contemporary data-image convertibility can come from any signal picked up by the range of instruments I previously discussed, even from extra-human ranges. All of these phenomena are, scientifically speaking, *wave phenomena*—radio waves are long, gamma rays are short. Now, with some ingenuity in instrumentation, virtually any data also can be turned into acoustic phenomena.

There is a cold war tale that will illustrate one of these possibilities: "bugging" traditionally has been done by placing an electronic device inside of the room to be bugged—anyone fond of spy movies has seen such events. But what if it is too late to bug a room and you want to know what is being said inside? If this room has windows, particularly nice big ones with single panes, with the proper equipment, which includes a laser connected through computational devices, you can regard the windowpane as a surrogate "speaker diaphragm" that, however minutely, is picking up the waves or vibrations of the speakers inside. Your finely tuned laser picks up these vibrations, turns them into data, and then reproduces them as acoustic presentations.

But now to Felix Hess, a more benign artistic practicioner: some of his installations were analogs to my bugging example, in that he aimed sensors at whole series of apartment windows; since he was not interested in individual voices, he picked up the acoustic rhythms of the collective dwellers. He also transformed these patterns by time compressing them. I mean, a whole day—as in time-lapse photography—is compressed into a few minutes. Thus in the morning there is a lot of apartment activity, then a quiet time when most people are at work, followed by another very active time when the dwellers return home—a sort of life rhythm song as it were.[4]

Perhaps the most unusual acoustic event was something Hess discovered accidentally: he noticed on some of his recordings that there was a strange, very low sound wave pattern that came and went, and that he could not recognize from any of his intended targets. It took some time before he recognized that this pattern was clearly associated with weather patterns. It appeared when the barometer readings were going down and disappeared when they went up. Then, with the help of meteorologists, he discovered that what he was hearing turned out to be the echos of giant storms off the coast of Iceland, which were being picked up by his sensitive equipment in Holland! The process of time compression also *brings into the range of human hearing infrasound*, that is, sound that is too low for us to hear. Whale songs, now familiar, are for the most part sounded in the frequency range too low for us to hear. These songs must be *translated* into higher frequencies that we can hear, and that is what has been produced for the now-popular recordings. With acoustic technologies, just as with those exceeding the visual spectrum, for humans to hear either infrasound or ultrasound constructive acoustical imaging must be employed. And, once again, the reflexively discovered, implicit role of embodiment again appears: What we have here is the precise acoustic auditory parallel to the constructed imaging illustrated usually in adronomy and other sciences. In this case sound, infrasound, too low to be heard by us, is technologically transformed, *translated*, into a range that we can hear.

- Once again, the implicit role of embodiment appears. For science, or art, to be experienced, it must take into account human embodiment.
- If the phenomenon lies beyond our capacity, then only by being technologically transformed can it come into our range.
- As these two acoustic examples show, the phenomenon can be so mediated by the computational capacity to convert image (or sound) into data and back into transformed image (or sound.) Given the cultural proclivity of science practice, there has already been produced a great deal of visual imagery but very little of auditory production.
- As my Hess example showed, actually in a rather close parallel to early radio effects, a genuine scientific discovery was made concerning the earth sounds of Icelandic storms.

I do not want to suggest that science is entirely deaf, since there is a lot of acoustic research that has been and is going on. Infrasound research is actually employed in many marine contexts. Some of it is controversial, since there is evidence that the sending of loud, low frequency transmissions probably is harmful to marine mammals (whales, seals, walruses, etc.). But, as should now be expected, most sonar research is actually converted into *visual imagery.*

My final example comes from oceanic mapping.

Figure 4.4. Ocean Bottom Map

Here is an excellent, concluding example of instrumental phenomenological variations in practice:

- Surface wave patterns are visually imaged and computer averaged to give a macro-scale "map" of the ocean floor. Gravity is such that seamounts (mountains under the sea) actually produce bulges on the ocean surfaces that can be detected through satellite imaging and tomographic averaging processes.
- For more refined results, multiscan sonar technology uses low-range sound waves for better resolution.
- For the finest resolution, one goes to remotely towed optical scans.

The end result is a three-dimensional *visual map* of the ocean floor. To produce it, one uses instruments that are productive of instrumental phenomenological variations, which in the end produce a visual *gestalt*, or pattern, which we humans instantaneously can recognize. Throughout

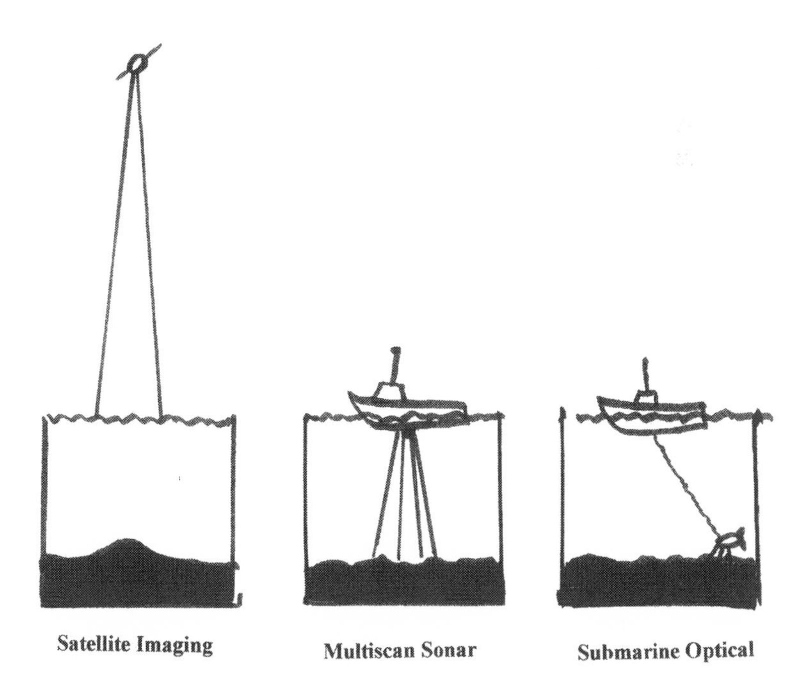

Figure 4.5. Mapping by Multiple Means

this process must run the critical, interpretive activity that I call material hermeneutics.

Finally, it may seem a bit strange that I began by making some suggestions about how the use of natural science instruments might well serve to change the ways in which we practice the humanities and the human sciences. Particularly in conjunction with historical disciplines, one would, at the least, enrich the narrative results; but in other cases, very different narratives are likely to be necessary.

Yet in the end I turned back to a science example in ocean mapping. What ties all of this together, however, is exactly the materiality of the hermeneutic process I have been suggesting. If such a hermeneutics is not entirely the *same* for traditional natural science and the human sciences, then it should by now be seen as continuous rather than disjunctive. For the contemporary world, that which had not been visible can now become visible, and that which was unheard can now begin to be heard. Things, too, have or may be given voices.

Notes

Chapter 1

1. John Dewey, *The Philosophy of John Dewey*, ed. John J. McDermott (New York: Putnam, 1977), 42.

2. Richard Rorty, *Consequences of Pragmatism* (Princeton, NJ: Princeton University Press, 1982), 197.

3. Dewey, *The Philosophy of John Dewey*, 161–62.

4. Ibid., 61.

5. Ibid, 61.

6. Hans Achterhuis, *American Philosophy of Technology: The Empirical Turn* (Bloomington: Indiana University Press, 2001), 6–8.

7. Ibid., 6–8.

8. Ibid.

Chapter 2

1. Carl Mitcham, *Thinking through Technology* (Chicago, IL: University of Chicago Press, 1994), 20.

2. Francis Bacon, *Novum Organum*, 1620.

3. Don Ihde, *Technics and Praxis: A Philosophy of Technology* (Dordrecht: Reidel, 1979).

4. Edmund Husserl, *The Crisis of European Sciences and Transcendental Phenomenology*, trans. David Carr (Evanston, IL: Northwestern University Press, 1970), 353.

5. Ibid., 370.

6. Ibid., 375.

7. Ibid., 70, emphasis.

8. Carl Mitcham, "From Phenomenology to Pragmatism," in *Postphenomenology: A Critical Companion to Ihde*, ed. Evan Selinger (Albany: State University of New York Press, 2006), 22.

9. Dick Teresi, *Lost Discoveries* (New York: Simon and Schuster, 2002), 38.

10. Ibid., 17.

11. I make a similar claim in "Husserl's Galileo Needed a Telescope," paper delivered at the Husserl Circle, Lima, Peru, 2003.

12. Over the years I have published a number of papers on Heidegger, as well as chapters in various books. *Technics and Praxis* (1979) includes my early work interpreting Heidegger's philosophy of technology; by the time of *Technology and the Lifeworld* (1990), I had become critical of aspects of Heidegger's analyses of technology and chapter eight contains a somewhat satirical caricature of that style of analysis. My "Deromanticizing Heidegger" article may be found in *Postphenomenology: Essays in the Postmodern Context* (Evanston, IL: Northwestern University Press, 1993).

13. Martin Heidegger, *Being and Time*, trans. E. Robinson and J. Macquarrie (New York: Harper and Row, 1962), 95.

14. Ibid., 97.

15. Ibid., 95.

16. Ibid., 102.

17. Ibid., 166, 181.

18. Ibid., 102.

19. While I agree that breakdown can produce such results, I disagree that this is the only way to discover the context, involvements, and relationships of technologies. One can discover the same phenomena through the application of phenomenological variations.

20. Martin Heidegger, *The Question Concerning Technology and Other Essays*, trans. W. Lovitt (New York: Harper Torchbooks, 1977), 304.

21. Historians of science and technology have more often held this view than philosophers of science. John Dewey was an early exception and, as noted previously, also held such a view.

22. Heidegger, *The Question Concerning* Technology, 245–46.

23. Ibid.

24. Ibid., 304–305.

25. Maurice Merleau-Ponty, *The Phenomenology of Perception*, trans. C. Smith (New York: Humanities Press, 1962), 250.

26. Ibid., 250.

27. Ibid., 143.

28. Ibid., 144.

29. Evan Selinger, who is working on technology transfers to developing countries, has pointed out how such technologies are usually seen to be positive, desirable, and enabling in contrast to older machine technologies.

30. Heidegger, *The Question*, 295.

31. Ibid., 294.

32. Internet citation, "The Restless Universe," www.physicalworld.org/restlessuniverse.

Chapter 3

1. See Don Ihde, *Technics and Praxis: A Philosophy of Technology* (Dordrecht: Reidel, 1979), and *Technology and the Lifeworld* (Bloomington: Indiana University Press, 1990).

2. Alfred North Whitehead, *Science and the Modern World* (New York: New American Library, 1963), 107.

3. See Don Ihde, *Instrumental Realism* (Bloomington: Indiana University Press, 1991). This book develops a praxis and perception theory of science.

4. David C. Lindberg, *The Beginnings of Western Science* (Chicago, IL: University of Chicago Press, 1992), 41.

5. See Robert Crease, *The Prism and the Pendulum: Science's Ten Most Beautiful Experiments* (New York: Random House, 2003).

6. Nigel Henbest and Michael Marten, *The New Astronomy* (Cambridge: Cambridge University Press, 1996), 6.

7. New manuscripts of Kepler's astrological casts were widely reported upon in 2000.

8. See Anthony Aveni, *Stairways to the Stars: Skywatching in Three Great Ancient Cultures* (New York: Wiley, 1997).

9. Daniel Boorstin, *The Discoverers* (New York: Vintage Books, 1983), 314–15.

10. See Dick Teresi, *Lost Discoveries* (New York: Simon and Schuster, 2002).

11. Edward R. Tufte, *Visual Explanations* (Cheshire: Graphics Press, 1997), 24.

12. Henbest and Marter, *The New Astronomy,* 6.

13. J. B. Hearnshaw, *The Analysis of Starlight* (Cambridge: Cambridge University Press, 1986).

14. Chen Cheng-Yih, "Acoustics in Chinese Culture," *Encyclopedia of the History of Science, Technology and Medicine in Non-Western Cultures* (Dordrecht: Kluwer, 1997), 10.

15. George Johnson, *Miss Leavitt's Stars: The Untold Story of the Woman Who Discovered How to Measure the Universe* (New York: Norton, 2005).

16. Edward R. Tufte, *Visual Explanation* (Cheshire Graphics Press, 1997), 24.

17. Betty Ann Kevles, *Naked to the Bone* (Reading, MA: Addison-Wesley, 1998).

18. Thomas Kuhn, *The Structure of Scientific Revolutions* (Chicago, IL: University of Chicago Press, 1962), 111.

19. See Bas Van Fraasen, *The Scientific Image* (Oxford: Clarendon Press, 1980).

Chapter 4

1. Don Ihde, "More Material Hermeneutics," *Yearbook of The Institute for Advanced Study on Science, Technology and Society* (Graz: Profil Verlag, 2004), 341–50.

2. Don Ihde, "Technologies-Musics-Embodiments," *Janus Head,* 10:1 (2007) 7–24.

3. Ibid.

4. Felix Hess, *Light as Air* (Heidelberg: Keher Verlag, 2003).

Selected Bibliography

1. Books

On Non-Foundational Phenomenology. Publicatioiner frän institutionen for pedagogic, Fenomenolografiska nötiser 3. Göteborg, 1986. This monograph published a set of lectures given in 1984 at Göteborg University, Sweden. The lectures were a response to pragmatism, particularly the neopragmatism of Richard Rorty, in which I adapt an antiessentialist, nonfoundational perspective for phenomenology.

Postphenomenology: Essays in the Postmodern Context. Evanston, IL: Northwestern University Press, 1993. Here, addressing themes of postmodernist thinkers, I use the term *postphenomenology* for the first time in a book title.

Expanding Hermeneutics: Visualism in Science. Evanston, IL: Northwestern University Press, 1998. This book concentrates on a reframing of the philosophy of science in hermeneutic terms and outlines a program related to imaging technology instrumentation in science.

Bodies in Technology. Minneapolis: University of Minnesota Press, 2002. Dialogues regarding the philosophy of science, philosophy of technology, and science studies, with an emphasis on the role of embodiment.

Chasing Technoscience: Matrix of Materiality. Edited with Evan Selinger. Bloomington: Indiana University Press, 2003. An examination of technoscience thinkers who emphasize the role of materiality in the practices of technoscience.

Human Studies. 31/1, 2008, *Postphenomenology Research Issue*, Ed. Don Ihde. Springer Publishers.

2. Articles Relevant to Chapters, Selected Only from 2000 on

Chapter 1

"Postphenomenology—Again?" *Working Paper No. 3.* Center for STS Studies, Aarhus University, Denmark, 2003, 3–25.

"Postphenomenology and the Lifeworld." *Phenomenology and Ecology, Twenty-third Annual Symposium of the Simon Silverman Phenomenology Center*, Duquesne University, 2006, 39–52.

Chapter 2

"Technology and the 'Other' Continental Philosophy." *Continental Philosophy Review*, 33 (Spring 2000): 59–74.

"Has the Philosophy of Technology Arrived?: A State-of-the-Art Review. *Philosophy of Science*, 71 (2004): 117–31.

"Philosophy of Technology." In *World and Worldhood*, ed. by Peter Kemp. Dordrecht: Springer, 2004, 91–107.

Chapter 3

"Visualism in Science." In *Visual Information Processing*. ed. Salvatore Soraci and Kymio Murata-Soraci. Westbrook: Praeger, 2002, 249–60.

"Simulation and Embodiment." *Yearbook of the Institute of Advanced Study on Science, Technology and Society*. Ed. Armo Bamme, Günter Getzinger, Bernhardt Wieser. Graz: Profil Verlag, 2004, 231–44.

"Models, Models Everywhere." In *Simulation: Pragmatic Construction of Reality, Sociology of the Sciences Yearbook*, ed. Johannes Lenhard, Günter Kuppers, and Terry Shin. Dordrecht: Springer, 2006, 79–88.

"Art Precedes Science: Or Did the *Camera Obscura* Invent Modern Science?" In *Mediated Vision*, ed. Petran Kockelkoren. Art EZ Press, 2007, 25–37.

"Hermeneutics and the New Vision." In *Ways of Seeing, Ways of Speaking*, ed. Kristie Fleckenstein, Sue Hum, and Linda Calendrillo. Chicago, IL: Parlor Press, 2007, 33–51.

Chapter 4

"More Material Hermeneutics." *Yearbook of the Institute of Advanced Study on Science, Technology and Society*. ed. Arno Bammé, Günter Getzmyer, Bernhardt Wieser. Graz: Profil Verlag, 2005, 341–50.

Index

Made in United States
Orlando, FL
20 February 2022

15011977R00062